T889
Problem solving and improvement: quality and other approaches

GW00725885

The Open University

Block 4 Methods and approaches

Block 5 Managing problem solving and improvement

Index of techniques

This publication forms part of the Open University module T889 *Problem solving and improvement: quality and other approaches*. Details of this and other Open University modules can be obtained from the Student Registration and Enquiry Service, The Open University, PO Box 197, Milton Keynes MK7 6BJ, United Kingdom (tel. +44 (0)845 300 60 90, email general-enquiries@open.ac.uk).

Alternatively, you may visit the Open University website at http://www.open.ac.uk where you can learn more about the wide range of modules and packs offered at all levels by The Open University.

To purchase a selection of Open University materials visit http://www.ouw.co.uk, or contact Open University Worldwide, Walton Hall, Milton Keynes MK7 6AA, United Kingdom for a brochure (tel. +44 (0)1908 858793; fax +44 (0)1908 858787; email ouw-customer-services@open.ac.uk).

The Open University
Walton Hall, Milton Keynes
MK7 6AA

First published 2007. Second edition 2011.

Edited and designed by The Open University.

Typeset by SR Nova Pvt. Ltd, Bangalore, India.

Printed and bound in the United Kingdom by Hobbs The Printers Limited, Brunel Road, Totton, Hampshire SO40 3WX.

ISBN 978 1 8487 3232 2

2.1

Block 4 Methods and approaches

CONTENTS

AIMS

The aims of Block 4 are to:

- introduce you to a variety of approaches to problem solving and improvement
- encourage you to think critically about the strengths and weaknesses of these approaches
- enable you to develop sufficient level of competence to start using these approaches in real-life situations where problem solving and improvement are required.

LEARNING OUTCOMES

After studying Block 4 you should be able to:

- start using a variety of approaches to problem solving and improvement in real-life situations
- demonstrate awareness of the strengths and weaknesses of a variety of approaches to problem solving and improvement
- for an application of one of the approaches and using a way that is appropriate to that approach, conduct an analysis and present the analysis and the findings.

1 INTRODUCTION

At the end of Block 1 you were introduced to three generic problem-solving and improvement methods based on three different metaphors for problem solving:

1 a learning cycle

2 a journey

3 a search.

Block 3 contained a large number of techniques that can be used to investigate and analyse, generate and implement recommendations, and monitor future performance. Most of those techniques can be used in two ways: in an ad hoc fashion to fulfil a particular need to investigate, present information, analyse or whatever; or as part of an organised approach to problem solving and improvement. For the reasons rehearsed in Block 1, the latter has particular strengths and is therefore something I shall consider further in this block, which describes a series of more specialised and more sophisticated approaches. Two of these approaches were designed by quality organisations, three draw on systems thinking, and two resulted from industry-led initiatives. The first two approaches – the ISO 9000 series approach and the Excellence approach – are very closely related to Total Quality Management (TQM), so in order to set the scene I shall summarise the key aspects of TQM before looking at the approaches.

According to a glossary of quality that appeared in the journal *Quality Progress* (1992), the term 'total quality management' was introduced in 1985 by the US Naval Air Systems Command to describe its Japanese-style management approach to quality improvement. Whatever the truth of this statement, it is clear that TQM developed from a range of teachings, including those of Crosby, Deming, Feigenbaum, Ishikawa, Juran and Taguchi.

The original key aspects of TQM were few in number. The first was continuous improvement. As you might guess, this incorporated a change to a more proactive attitude which sought improvements in situations where no obvious symptoms of problems existed. The second was the formation of multifunctional teams, for design, improvement, and so on, and its inevitable effect of breaking down organisational barriers. The third aspect was reduction in variation, and the fourth was supplier integration. The final aspect was the use of education and training not only to facilitate the use of techniques but also for their motivating effects.

If you were to study the quality literature of the late 1980s and early 1990s it would soon become clear that TQM has always meant different things to different people. For some it was a shorthand way of denoting a modern, effective way of managing quality whereas for others it was a specific approach, and some people even portrayed it as a package that could be

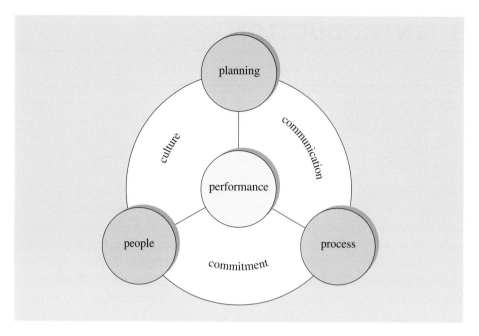

Figure 4.1   The new framework for Total Quality Management (Source: Oakland, 2005, p. 1059)

bought off the shelf. (It was this portrayal as a package that was most responsible for TQM being denounced as a quick fix because many of the companies that called in consultants to 'install TQM' quickly abandoned it.)

It is not relevant here to study in detail all the forms that TQM has taken but it is worth looking at one version that Oakland (2005) calls 'the new framework for Total Quality Management' because it forms a useful backdrop for studying the next two sections of this block. Figure 4.1 shows the new framework and Box 4.1 contains Oakland's description of it.

BOX 4.1   THE NEW FRAMEWORK FOR TOTAL QUALITY MANAGEMENT

The methods underlying this framework offer the detail required for making improvements in performance – a prescriptive approach that many organizations find helpful in these days of complicated management theory.

... *processes* are the key to delivering quality of products and services to customers ... [and] *processes* are a key linkage between the enablers of *planning* (leadership driving policy and strategy, partnerships and resources), through *people* into the *performance* regarding customers, people, society and key outcomes. These four Ps are vital to delivering quality products and services to customers, and form a structure of 'hard management necessities' for the new simple TQM model.

Planning – includes the development and deployment of policies and strategies; setting up appropriate partnerships and resources; and designing in quality.

Performance – includes establishing a performance measurement framework – a 'balanced scorecard' for the organization; carrying out self-assessment, audits, reviews and benchmarking.

Processes – includes process understanding, management, design and re-design; quality management systems; continuous improvement.

People – includes managing the human resources; culture change; teamwork; communications; innovation and learning.

Wrapping around all this to ensure successful implementation is, of course, effective leadership and commitment.

The core of this new model needs to be surrounded by *commitment* to quality and meeting the customer requirements, *communication* of the quality message, and recognition of the need to change the *culture* of most organizations to achieve total quality. These are the 'soft foundations' that must host the hard necessities of planning, people and processes.

The key to successful implementation is about learning how to manage in a total quality way using this structure.

(Oakland, 2005, pp. 1058–1060)

The next section looks at the first of two approaches that are underpinned by TQM – the ISO 9000 series.

2   THE ISO 9000 SERIES APPROACH

The International Organization for Standardization (ISO) is a network of the national standards institutes of 158 countries and is the largest developer and publisher of International Standards in the world. It claims to enable 'a consensus to be reached on solutions that meet both the requirements of business and the broader needs of society' (ISO, 2010a).

The vast majority of ISO standards (there are over 17 000 of them) are specific to a particular product, material, or process but ISO 9001 and its sister series ISO 14000, are known as 'generic management system standards'. (The ISO 14000 series is primarily concerned with the environmental management of an organisation in terms of minimising its harmful effects on the environment and continually improving its environmental performance.) ISO explains the term 'generic management standards' thus:

> 'Generic' means that the same standard can be applied to any organization, large or small, whatever its product or service, in any sector of activity, and whether it is a business enterprise, a public administration, or a government department.
>
> (ISO, 2010b)

2.1   The ISO 9000 series

The international standard ISO 9000 began life as the British standard BS 5750 *Quality Systems*. BS 5750 made its first appearance in 1979. It interpreted quality in the sense of 'fitness for purpose' and 'safe in use' and, according to the executives guide to it (Department of Trade and Industry (DTI), 1992), was designed to 'tell suppliers and manufacturers what is required of a quality-orientated system' (p. 4) and set out 'how you can establish, document and maintain an effective quality system which will demonstrate to your customer that you are committed to quality and are able to supply their "quality needs"' (p. 2). It was revised in 1987, became dual numbered as international standard ISO 9000 and further revised in 1994 when ISO 9004-4:1993 *Total Quality Management – Guidelines for Quality Improvement* was added. However, it is fair to say that at this stage the main thrust of ISO 9000 was still quality assurance rather than prevention and improvement, and its main attraction was the certification scheme that it underpinned.

In 1997 a large global survey of 1120 users and customers of the standard was conducted in order to inform the next revision process. A series of needs was identified, three of which are particularly relevant here:

- The revised standards should have a common structure based on a Process model
- Requirements should include demonstration of continuous improvement and prevention of non-conformity

- The revised standards should facilitate self-evaluation

(ISO, 1998, pp. 1–2)

Revised standards that were designed to meet the needs identified were officially released by ISO on 15 December 2000 and were given the overall title of the ISO 9000 series or family of standards. They were published in the UK as:

BS EN ISO 9000:2000 *Quality management systems. Fundamentals and vocabulary*

BS EN ISO 9001:2000 *Quality management systems. Requirements*

BS EN ISO 9004:2000 *Quality management systems. Guidelines for performance improvements*

These have since been revised and the latest versions are:

BS EN ISO 9000:2005 *Quality management systems. Fundamentals and vocabulary*

BS EN ISO 9001:2008 *Quality management systems. Requirements*

BS EN ISO 9004:2009 *Managing for the sustained success of an organization. A quality management approach*

As time has gone on the emphasis on continual improvement within the framework of quality systems has increased significantly. Continual improvement is now set out as one of the eight quality management principles on which the standards are based and many of the means of bringing about improvement are enshrined in the other principles:

a) **Customer focus**
    Organizations depend on their customers and therefore should understand current and future customer needs, should meet customer requirements and strive to exceed customer expectations.

b) **Leadership**
    Leaders establish unity of purpose and direction of the organization. They should create and maintain the internal environment in which people can become fully involved in achieving the organizations objectives.

c) **Involvement of people**
    People at all levels are the essence of an organization and their full involvement enables their abilities to be used for the organizations benefit.

d) **Process approach**
    A desired result is achieved more efficiently when activities and related resources are managed as a process.

e) **System approach to management**
    Identifying, understanding and managing interrelated processes as a system contributes to the organizations effectiveness and efficiency in achieving its objectives.

f) **Continual improvement**
    Continual improvement of the organization's overall performance should be a permanent objective of the organization.

g) **Factual approach to decision making**
    Effective decisions are based on the analysis of data and information.

h) **Mutually beneficial supplier relationships**
    An organization and its suppliers are interdependent and a mutually beneficial relationship enhances the ability of both to create value.

(BS EN ISO 9000:2005, pp. v–vi)

As you might expect, ISO is very precise in the terminology it uses. Quality improvement and continual improvement are both defined within the standard:

quality improvement

part of **quality management** focused on increasing the ability to fulfil quality requirements

NOTE The requirements can be related to any aspect such as **effectiveness, efficiency** or **traceability**.

(BS EN ISO 9000:2005, clause 3.2.12, p. 9)

continual improvement

recurring activity to increase the ability to fulfil **requirements**

NOTE The **process** of establishing objectives and finding opportunities for improvement is a continual process through the use of **audit findings** and **audit conclusions**, analysis of data, management **reviews** or other means and generally leads to **corrective** or **preventive action**.

(BS EN ISO 9000:2005, clause 3.2.13, p. 9)

ISO 9001:2008 sets out the requirements for compliance to the standard. It is based on what it calls 'a model of a process-based quality management system' where improvement is seen as part of the system, sitting alongside measurement and analysis, and the need continually to improve the quality management system itself is also identified. The standard spells out this requirement thus:

The organization shall continually improve the effectiveness of the quality management system through the use of the quality policy, quality objectives, audit results, analysis of data, corrective and preventive actions and management review.

(BS EN ISO 9001:2008, p. 14)

ISO 9004:2009 expands this model considerably in order to provide 'a wider focus on quality management than ISO 9001' (BS EN ISO 9004:2009, p. v). The extended version is shown in Figure 4.2.

The purpose of ISO 9004:2009 is set out as being 'to support the achievement of sustained success for any organization in a complex, demanding, and ever-changing environment, by a quality management approach' and it promotes 'self-assessment as an important tool for the review of the maturity level of the organization' (BS EN ISO 9004:2009, p. v). The self-assessment tool itself is provided in Annex A of the standard (pp. 20–37). An extract from that Annex is shown in Box 4.2.

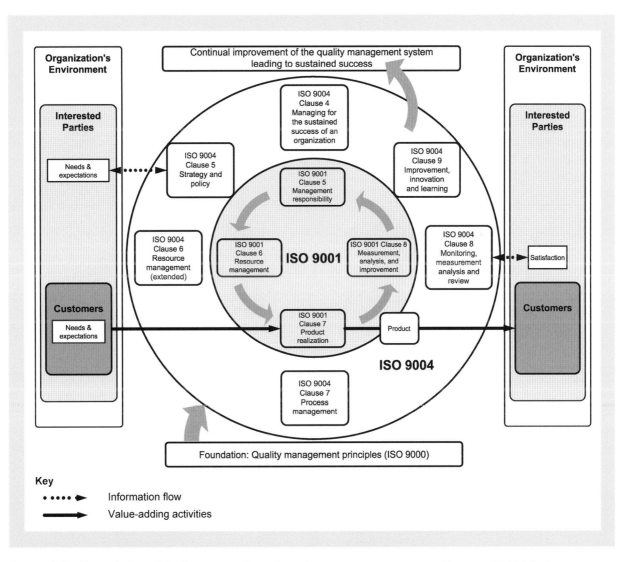

Figure 4.2   Extended model of a process-based quality management system (Source: BS EN ISO 9004:2009, p. v)

BOX 4.2   A SECTION OF THE SELF-ASSESSMENT TOOL

Key element	Level 1	Level 2	Maturity level Level 3	Level 4	Level 5
How are results achieved? (Monitoring and measurement)	Results are achieved in a random manner. Corrective actions are ad hoc.	Some predicted results are achieved. Corrective and preventive actions are performed in a systematic way.	Predicted results are achieved, especially for identified interested parties. There is consistent use of monitoring, measurement and improvement.	There are consistent, positive, predicted results, with sustainable trends. Improvements and innovations are performed in a systematic way.	The achieved results are above the sector average for the organisation, and are maintained in the long term. There is implementation of improvement and innovation throughout the organisation.
How are results monitored? (Monitoring and measurement)	Financial, commercial and productivity indicators are in place.	Customer satisfaction, key realisation processes and the performance of suppliers are monitored.	The satisfaction of the organisation's people and its interested parties is monitored.	Key performance indicators are aligned with the organisation's strategy and are used for monitoring.	Key performance indicators are integrated into the real-time monitoring of all processes, and performance is efficiently communicated to relevant interested parties.
How are improvement priorities decided? (Improvement, innovation and learning)	Improvement priorities are based on errors, complaints or financial criteria.	Improvement priorities are based on customer satisfaction data, or corrective and preventive actions.	Improvement priorities are based on the needs and expectations of some interested parties, as well as those of suppliers and the organisation's people.	Improvement priorities are based on trends and inputs from other interested parties as well as analysis of social, environmental and economic changes.	Improvement priorities are based on inputs from emerging interested parties.
How does learning occur? (Improvement, innovation and learning)	Learning occurs randomly, at an individual level.	There is systematic learning from the organisation's successes and failures.	A systematic and shared learning process is implemented in the organisation.	There is a culture of learning and sharing in the organisation that is harnessed for continual improvement.	The organisation's processes for learning are shared with relevant interested parties, and support creativity and innovation.

NOTE: The current maturity level of the organisation's individual elements is the highest level achieved up to that point with no preceding gaps in the criteria.
Source: BS EN ISO 9004:2009, p. 25.

ACTIVITY 4.1 . . . .

Given that Level 1 for Partners and suppliers is 'Supplier communication are limited to tendering, order placement or problem resolution' and Maturity Level 5 is 'Data demonstrates that partners are engaged in and are contributing to the organizations successes' what do you suggest levels two, three and four might be? ●

2.2   Evaluating the ISO 9000 series approach

There is very little to criticise in each of the separate parts of the 9000 series of standards that make up this approach. ISO 9004:2009 recommends the P–D–C–A cycle, which you saw in Block 1, and many other 'universal goods' such as the effective and efficient use of resources, a 'planned, transparent, ethical and socially responsible approach' (p. 6) to the management of people and the need to plan and control processes. However, in its concentration on the need for self-assessment the current version of the 9004 aligns itself very strongly with the Excellence approach you will be meeting in the next section. And therein lies a problem. ISO 9001:2000 and ISO 9004:2000 were developed as a 'consistent pair' of standards having the same structure and sequence with the aim of facilitating 'an easy and useful transition between them'. ISO advised organisations to adopt 9001 'to achieve a first level of performance' and then move on to 9004 'to make your quality management system increasingly effective in achieving your own business goals' (ISO, n.d.). The BS catalogue described ISO 9004:2000 as providing guidelines beyond the requirements of ISO 9001. Although BS EN ISO 9000:2000 was withdrawn in 2006 and replaced by BS EN ISO 9000:2005 and BS EN ISO 9001:2000 was replaced by BS EN ISO 9001:2008 the changes were minor in both cases so the consistency between the standards in the series remained. But BS EN ISO 9004:2009 is very different from ISO 9004:2000. The consistency between 9001 and 9004 has been lost. ISO 9004 used to take each of the requirements of ISO 9001 and offer specific clause-by-clause advice on how they could be met; it was thus an implementation guide in a very real sense. Not only does it no longer fulfill that role, it also deals with topics that are not included in ISO 9001. These include:

strategy and policy formulation

strategy and policy deployment

financial resources

knowledge, information and technology

natural resources

innovation and learning.

The overall result of the changes is that the ISO 9000 series can still be classed as an approach but it is one that has a certain lack of coherence within it.

The three standards (9000, 9001 and 9004) are available via 'British standards online' in the Databases section of the OU library website. It is particularly important you look at them if you are considering using the ISO 9000 series approach in your project.

3   THE EXCELLENCE APPROACH

3.1   The Excellence Model

The European Foundation for Quality Management (EFQM) was formed in 1988 and officially established in 1989. One of its first strategic objectives was to create a European Quality Award along the lines of the Malcolm Baldrige Award which had been competed for in the USA since 1987. Following the pattern of Baldrige, two things were needed to launch the award: a model of quality and a detailed assessment procedure. The former appeared as the European Model for Business Excellence in time for the first ever European Quality Award in 1992. In 2000 the model was modified and renamed as the EFQM Excellence Model, and was then modified again in 2003. The version that was launched in 2010 is shown in Figure 4.3.

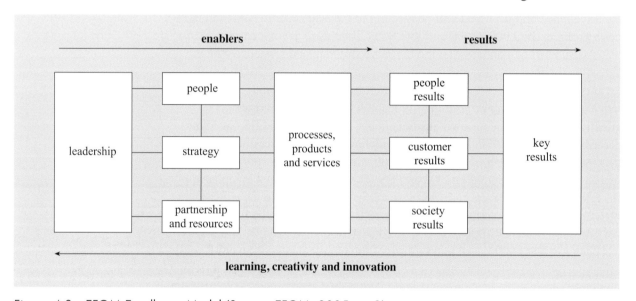

Figure 4.3   EFQM Excellence Model (Source: EFQM, 2005, p. 9)

The concepts that underpin the model are as follows:

Achieving Balanced Results
Excellent organisations meet their Mission and progress towards their Vision through planning and achieving a balanced set of results that meet both the short and long term needs of their stakeholders and, where relevant, exceed them.

Adding Value for Customers
Excellent organisations know that customers are their primary reason for being and strive to innovate and create value for them by understanding and anticipating their needs and expectations.

Leading with Vision, Inspiration and Integrity
Excellent organisations have leaders who shape the future and
make it happen, acting as role models for its Values and ethics.

Management by Processes
Excellent organisations are managed through structured and
strategically aligned processes using fact-based decision making to
create balanced and sustained results.

Succeeding through People
Excellent organisations value their people and create a culture of
empowerment for the balanced achievement of organisational and
personal goals.

Nurturing Creativity and Innovation
Excellent organisations generate increased value and levels of
performance through continual and systematic innovation by
harnessing the creativity of their stakeholders.

Building Partnerships
Excellent organisations seek, develop and maintain trusting
relationships with various partners to ensure mutual success. These
partnerships may be formed with customers, society, key suppliers,
educational bodies or Non-Governmental Organisations (NGOs).

Taking Responsibility for a Sustainable Future
Excellent organisations embed within their culture an ethical
mindset, clear Values and the highest standards for organisational
behaviour, all of which enable them to strive for economic, social
and ecological sustainability.

(EFQM, 2009a)

If you turn back to the eight quality management principles on which the
ISO 9000 series of standards is based you will see a large amount of overlap
between them and the concepts underpinning the EFQM Excellence Model.

ACTIVITY 4.2 . . . .

I would suggest that in general terms the concepts underpinning the EFQM
Excellence Model are broader than the eight quality management principles
on which the ISO 9000 series is based. Identify two instances where this
appears to be the case. ●

The blocks of the model represent nine criteria against which an
organisation's progress towards excellence can be judged. Five of the criteria
are concerned with what the organisation does and are called 'enablers', and
the other four cover what the organisation achieves and are classed as
'results'. 'Learning, creativity and innovation' is a common theme across the
whole.

EFQM describes each of the criteria as follows, giving first the definition and then the sub-criteria:

1. Leadership

Excellent organisations have leaders who shape the future and make it happen, acting as role models for its values and ethics and inspiring trust at all times. They are flexible, enabling the organisation to anticipate and react in a timely manner to ensure the ongoing success of the organisation.

1a. Leaders develop the Mission, Vision, Values and ethics and act as role models

1b. Leaders define, monitor, review and drive the improvement of the organisation's management system and performance

1c. Leaders engage with external stakeholders

1d. Leaders reinforce a culture of excellence with the organisation's people

1e. Leaders ensure that the organisation is flexible and manages change effectively

2. Strategy

Excellent organisations implement their Mission and Vision by developing a stakeholder focused strategy. Policies, plans, objectives and processes are developed and deployed to deliver the strategy.

2a. Strategy is based on understanding the needs and expectations of both stakeholders and the external environment

2b. Strategy is based on understanding internal performance and capabilities

2c. Strategy and supporting policies are developed, reviewed and updated

2d. Strategy and supporting policies are communicated, implemented and monitored

3. People

Excellent organisations value their people and create a culture that allows the mutually beneficial achievement of organisational and personal goals. They develop the capabilities of their people and promote fairness and equality. They care for, communicate, reward and recognise, in a way that motivates people, builds commitment and enables them to use their skills and knowledge for the benefit of the organisation.

3a. People plans support the organisation's strategy

3b. People's knowledge and capabilities are developed

3c. People are aligned, involved and empowered

3d. People communicate effectively throughout the organisation

3e. People are rewarded, recognised and cared for

4. Partnerships and Resources

Excellent organisations plan and manage external partnerships, suppliers and internal resources in order to support strategy and policies and the effective operation of processes. They ensure that they effectively manage their environmental and societal impact.

4a. Partners and suppliers are managed for sustainable benefit

4b. Finances are managed to secure sustained success

4c. Buildings, equipment, materials and natural resources are managed in a sustainable way

4d. Technology is managed to support the delivery of strategy

4e. Information and knowledge are managed to support effective decision making and to build the organisational capability

5. Processes, Products and Services

Excellent organisations design, manage and improve processes to generate increasing value for customers and other stakeholders.

5a. Processes are designed and managed to optimise stakeholder value

5b. Products and Services are developed to create optimum value for customers

5c. Products and Services are effectively promoted and marketed

5d. Products and Services are produced, delivered and managed

5e. Customer relationships are managed and enhanced

6. Customer Results

These are divided into two parts:

6a. Customers' perceptions of the organisation

6b. Internal measures ... to monitor, understand, predict and improve the performance of the organisation and to predict their impact on the perceptions of its external customers

7. People Results

These are also divided into:

7a. People's perceptions of the organisation

7b. Internal measures ... to monitor, understand, predict and improve the performance of the organisation's people and to predict their impact on perceptions

8. Society Results

These are also divided into:

8a. Society's perceptions of the organisation

8b. Internal measures ... to monitor, understand, predict and improve the performance of the organisation and to predict the impact on the perceptions of society

9. Key Performance Results

These are divided into:

9a. Key financial and non-financial outcomes

9b. Key financial and non-financial indicators that are used to measure the organisation's operational performance

(EFQM, 2009a)

3.2   Using the model

The Excellence Model has always been intended to be 'a tool to help drive improvement' (EFQM, 2009b). The main mechanism for driving this is self-assessment. The general steps involved in undertaking self-assessment are:

> planning
>
> data collection
>
> assessment
>
> identification of strengths and areas for improvement
>
> action planning.

There is thus a strong link between self-assessment and improvement.

EFQM recommends that when the Excellence Model is used for self-assessment the sub-criteria under each of the nine headings should be addressed using a 'logic' that it calls RADAR (results, approaches, deploy, assess and refine). This is represented in Figure 4.4.

You should now watch Programme 2: Recognising Excellence. The purpose of this programme is to give you a practical understanding of what it means to use the EFQM Excellence Model in the management of an organisation. This programme was made before the Model was revised for 2010 but its content remains equally valid today.

3.3   Evaluating the Excellence approach

The uses of the Excellence Model only go as far as the relatively early stage of problem solving and improvement: identification of opportunities. RADAR goes beyond that, but at a relatively high level of abstraction. Therefore, the biggest contribution the Excellence approach makes to problem solving is through the use of self-assessment. Oakland's reservations expressed in 2005 are still relevant today:

> Whilst the award models provide frameworks for understanding quality and excellence, their non-prescriptive nature makes them unsuitable for bridging the quality gap – they are not implementation models. A self-assessment process against these frameworks provides insights into the deficiencies in the organization's enablers and the gaps in actual performance, but it

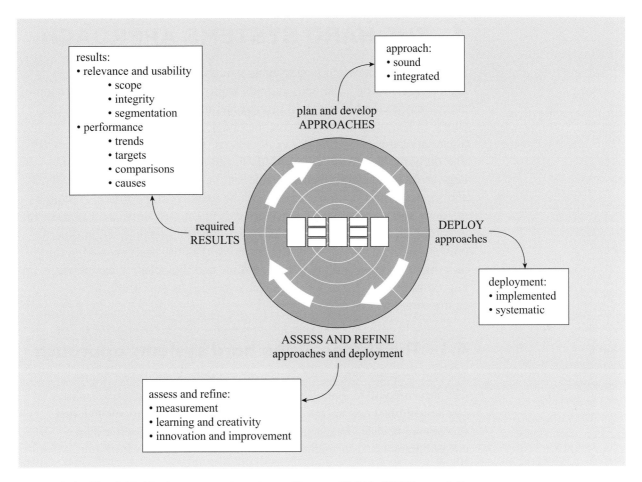

Figure 4.4   The RADAR elements and attributes (Source EFQM, 2009b, p. 14)

does not provide specific guidance on best practice and does not tell you how to bridge those gaps.

(Oakland, 2005, p. 1058)

Now read Offprint 7.

4   THE HARD SYSTEMS APPROACH

In this and the following two sections you will meet three systems approaches. The first, the hard systems approach, is a more formal version of the journey method of problem solving and improvement that you met in Block 1. The second, soft systems methodology, is an approach to organisational change that allows ill-defined, messy problems to be tackled. The third, the Systems Failures Method, can provide understanding of large-scale, systemic problems, enable learning from mistakes to take place and suggest ways in which failures can be avoided in the future. Although the three systems approaches are clearly different they do share a number of important characteristics that are absent or much less prominent in other problem-solving and improvement methods: they use the notion of system; they strive to be holistic; they are iterative; and they place great emphasis on the use of models. Look for these characteristics as you study the approaches.

4.1   The stages of the hard systems approach

The hard systems approach is designed to tackle problems or improvements that are amenable to quantification, when what would constitute success can be identified and agreed but where the steps needed to achieve that success are unclear. The stages of the hard systems approach are set out in Figure 4.5. You might like to compare this figure with Figure 1.5 in Block 1.

Now let us look at each of the stages in turn.

Stage 1: System description

This stage involves identifying and describing the problem to be solved (or the improvement to be made) and the existing 'system', including its environment and behaviour. A qualitative model of the existing state of affairs is developed using techniques such as systems mapping and multiple-cause diagramming, which you met in Block 3. Figures 4.6 and 4.7 are examples of a multiple-cause diagram and a systems map respectively. They relate to a problem situation where an agrochemical company, Company A, is in danger of losing market share for two reasons: first, because its main existing product, an aphicide, is losing some of its effectiveness and thus failing to satisfy customers' requirements; and, second, because its main competitor, Company C, is developing a better first-generation chemical. Company A is faced with the option of trying to develop a new first-generation pesticide of its own or developing new ways of keeping market share.

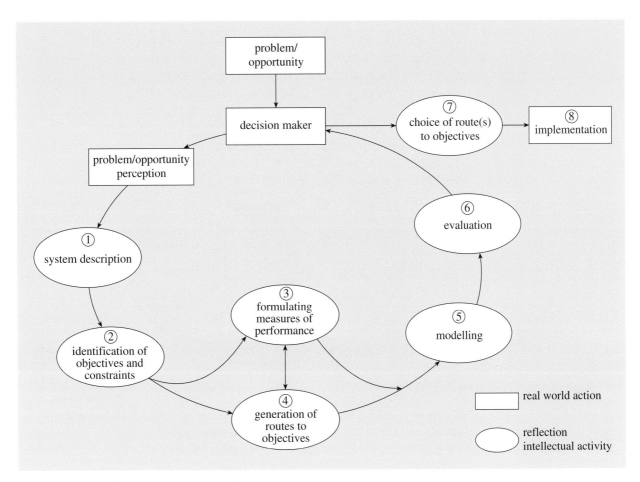

Figure 4.5   The hard systems approach

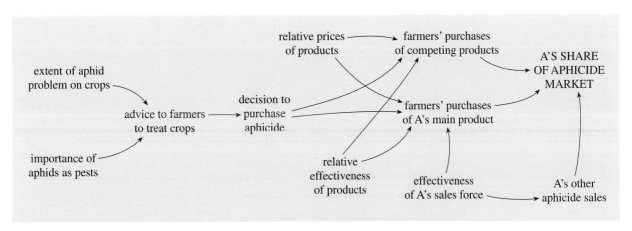

Figure 4.6   Multiple-cause diagram exploring A's share of aphicide market

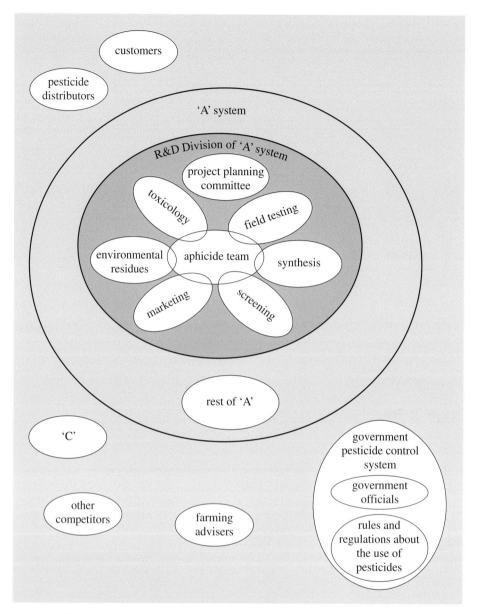

Figure 4.7   Systems map of A

Stage 2: Identification of objectives and constraints

Next, quantifiable and non-quantifiable objectives and constraints are
identified and then discussed so that the decision makers involved in the
problem or improvement can reach agreement. Usually, the objectives will
relate to variables that are inside the system boundary or could be brought
inside, while constraints relate to features in the environment. For example,
in the agrochemical problem an objective might be to maintain or increase
market share in Europe and a constraint might be the amount of money
available for investment in research and development. Figure 4.8 shows an

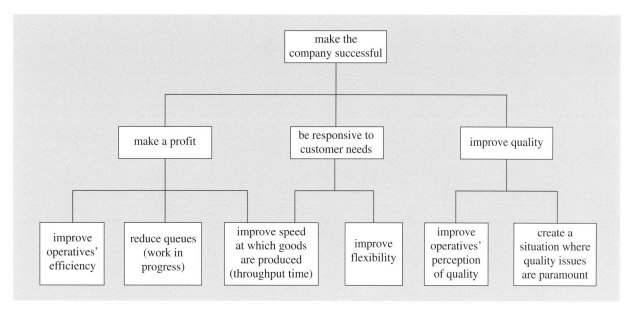

Figure 4.8   An objective tree

objective tree that might be appropriate for any commercial organisation losing market share.

Stage 3: Formulating measures of performance

The objectives are now used as a basis for deriving measurable means of assessing potential solutions in order to find the optimal route. Time to full product development and projected market share in five, seven and nine years' time would be among appropriate measures of performance in the agrochemical example.

Stage 4: Generation of routes to objectives

This stage entails generating a range of different ways of achieving the defined objectives and then narrowing them down to those which take account of the constraints and are feasible. For example, routes to the objective of maintaining or increasing market share in Europe might include the following:

1   Pursue rapid development of a new first-generation aphicide.
2   Develop new, effective second-generation aphicides.
3   Explore possibilities of entering into licensing arrangements.
4   Review past R&D work to see if potential products have been overlooked.

Stage 5: Modelling

Quantitative models that will predict the behaviour of the system when following particular routes to objectives are built and tested. A variety of

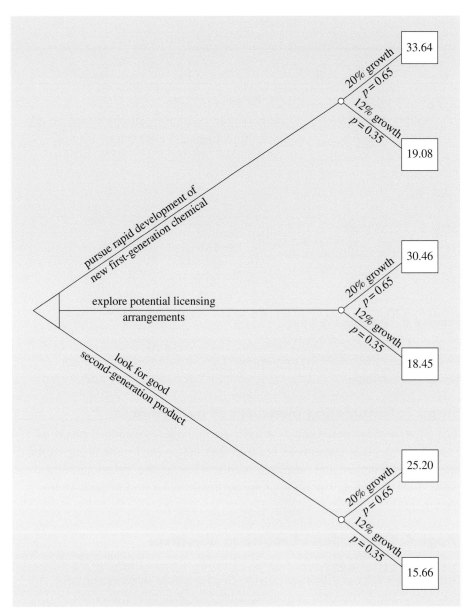

Figure 4.9   A decision tree for the agrochemical problem

techniques are available including simulation packages and spreadsheet modelling.

Stage 6: Evaluation

Using the models, each route is evaluated where possible against all the measures of performance. Decision trees can be a useful way of setting out the options. Figure 4.9 is an example of a decision tree for the agrochemical problem. In this figure there are three routes to objectives (that is, possible courses of action) and two possible outcomes for each route. The outcomes have been calculated assuming 12% and 20% growth and are expressed as net present values (£ million). The probability of 12% growth being achieved has been predicted as 0.35 by the people involved and the probability of 20% growth has been predicted at 0.65.

Stage 7: Choice of route(s) to objectives

It is unlikely that the same route is best in relation to all objectives, so all the results and the non-quantifiable objectives and constraints must be considered before making a final decision. The final choice is likely to involve combining routes that performed well against some measures but not others.

Stage 8: Implementation

This is the stage where the decision that has been made is translated into action.

4.2 Evaluating the hard systems approach

This section started by saying:

> The hard systems approach is designed to tackle problems or improvements that are amenable to quantification, when what would constitute success can be identified and agreed but where the steps needed to achieve that success are unclear.

Within these design limitations the approach can work extremely well. However, if objectives are unclear and there are few quantifiable measures of success or failure, then the hard systems approach is much less likely to lead to successful change, especially since it can assess only quantifiable alternative routes to objectives. Another potential drawback is that the sequential nature of the approach might lead to weak feedback from stage to stage.

The next section deals with an approach that is designed specifically for situations where objectives are unclear and there are few quantifiable measures of success or failure: the soft systems methodology.

5   SOFT SYSTEMS METHODOLOGY

Soft systems methodology (SSM) provides a means of tackling complex, ill-defined problems by redefining the problem content into a number of well-defined discussion points that will allow people to move towards positive action. It has been summarised as follows:

> SSM is a methodology that aims to bring about improvement in areas of social concern by activating in the people involved in the situation a learning cycle which is ideally *never-ending*. The learning takes place through the iterative process of using systems concepts to reflect upon and debate perceptions of the real world, taking action in the real world, and again reflecting on the happenings using systems concepts.
>
> The reflection and debate is structured by a number of systemic models. These are conceived as holistic ideal types of certain aspects of the problem situation rather than as accounts of it. It is taken as given that no objective and complete account of a problem situation can be provided.
>
> (von Bulow, 1989, pp. 35–6)

5.1   The stages of SSM

I shall describe here the original version of SSM (Checkland, 1972; 1981) as shown in Figure 4.10. For comparison, representations of the two later

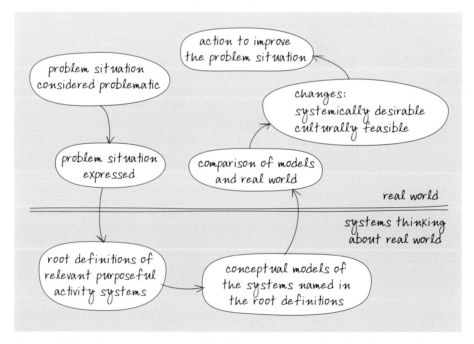

Figure 4.10   The seven-stage model of SSM (Source: Checkland and Scholes, 1990, p. 27)

versions are shown in Figures 4.11 and 4.12. I have selected the original version because the guidance available on how to use it is much more solid. I would suggest that the later versions rely on a level of craft skill and knowledge that is much more difficult for the non-specialist to build up, but if you want to pursue them further you might like to follow up the references given in the figure captions.

Stage 1: Problem situation considered problematic

Stage 1 begins, not with a 'problem' as such, but with a mess, or even with an organisational setting in which complex, ill-defined problems are simply thought to reside. (The term 'mess' was introduced in Box 1.2 of Block 1, and part of that box is reproduced here as Box 4.3 to remind you of its meaning.) Stage 1 also requires recognition that people involved in the

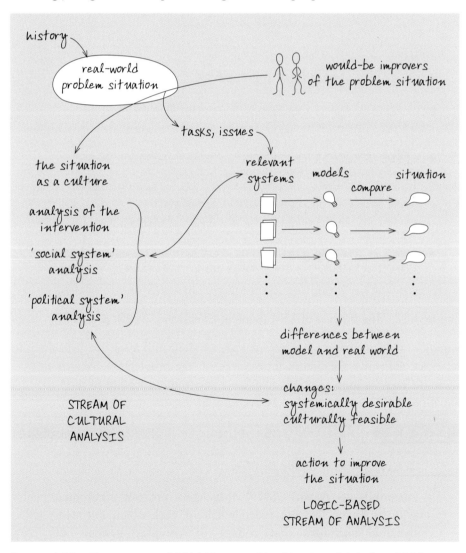

Figure 4.11   The process of SSM (Source: Checkland and Scholes, 1990, p. 29)

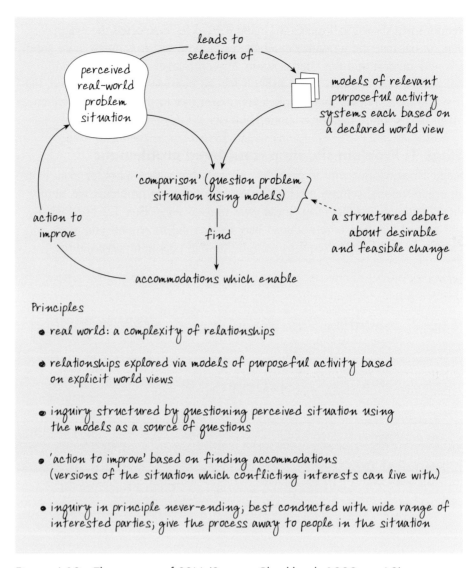

Principles

- real world: a complexity of relationships

- relationships explored via models of purposeful activity based on explicit world views

- inquiry structured by questioning perceived situation using the models as a source of questions

- 'action to improve' based on finding accommodations (versions of the situation which conflicting interests can live with)

- inquiry in principle never-ending; best conducted with wide range of interested parties; give the process away to people in the situation

Figure 4.12   The process of SSM (Source: Checkland, 1999, p. A9)

situation will inevitably have different views about it, views that lead them to advocate their own candidates for the role of 'the problem'. Furthermore, these differences in perceptions may be a determining feature of the situation, and therefore need to be recognised for that reason too.

BOX 4.3   DICHOTOMIES OF PROBLEMS

Ackoff ... messes versus problems

According to Ackoff (1979) 'Managers are not confronted with problems that are independent of each other, but with dynamic situations that consist of complex systems of changing problems that interact with each other. I call such situations messes. Problems are abstractions extracted from messes by analysis; they are to messes as atoms are to tables and chairs.' Individual problems may be 'solved'. But if they are components of a mess, the solutions to individual problems cannot be added, since those solutions will interact. Problems may be solved; messes need to be managed. If we insist on the solution mode, analysts will be relegated to those relatively minor problems which are nearly independent, while messes go inadequately managed (Ackoff, 1981).

(Source: T889, Block 1, p. 29)

Stage 2: Problem situation expressed

The technique used to present this information is the rich picture. An example of a rich picture of TNT Express's situation at the time Programme 2 was made is shown in Figure 4.13.

ACTIVITY 4.3 ·

Look again at Programme 1 and draw a rich picture of either the police situation before the introduction of statistical process control (SPC) or the police situation since the introduction of SPC. ●

Once the rich picture has been assembled, it is used to look for primary tasks and issues. Primary tasks are the tasks that the organisation in question was created to perform, or the tasks that an enterprise must perform if it is to survive. Searching for primary tasks is thus a way of finding an answer to the question: 'what is really central to this problem situation?'.

ACTIVITY 4.4 ·

Suggest two primary tasks for TNT Express. ●

In this context issues are general patterns or aspects that express or encapsulate important characteristics of the situation – that is, characteristics that are thought to be important in relation to the purpose of the investigation. The issues could be identified by looking at the topics or matters that are of concern or are the subject of dispute.

It is worth pausing here to look back at Figure 4.10. You will see that as you move from stage 2 to stage 3 you cross the double line that runs across it, separating the 'real world' from the 'abstract world' of systems thinking.

Stage 3: Root definitions

Stage 3 begins with a search for systemic ways of viewing the situation – ways that are articulated by naming hypothetical systems, known as relevant systems in SSM terminology. These are systems that are in some way

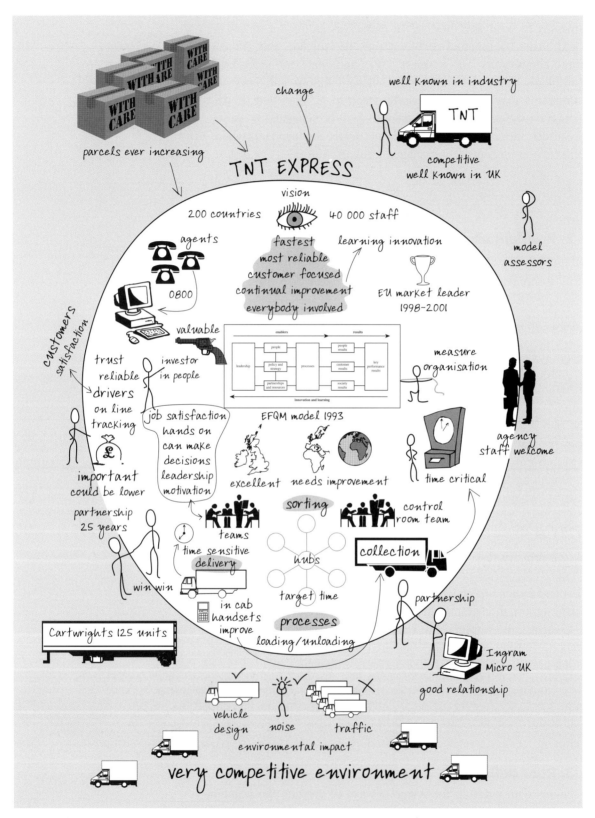

Figure 4.13   A rich picture of TNT Express's situation at the time Programme 2 was made

relevant to the problem situation in the sense that they will yield insight into the situation; they can be issue based or primary task based. For example, in the case of TNT Express a relevant system might be: a system to collect parcels from customers and deliver them in a timely fashion to their addressees. A narrower relevant system might be: a system to manage the collection and delivery of parcels.

A question that many newcomers to this approach ask when they encounter the concept of a relevant system is: 'relevant to what?'. The answer is: 'relevant to the process of improving the problem situation.' Clearly, you cannot be sure that a particular way of looking at a situation in systemic terms will turn out to be useful in the end, but you have to start somewhere and rely on the process of iteration to move you forward if you seem to be going nowhere or reach a dead end. However, I can assure you that people usually know straight away when a really insightful relevant system has been suggested, simply because of the unexpected perspective it provides.

It is very important to remember that a relevant system is not a system to 'solve' the problems inherent in the situation from which it stems, nor is it a system that anyone is ever going to design and implement in the real world. Its function is simply to provide an alternative way of viewing the problem situation; when it has been developed further in succeeding stages of the approach, it can then be compared and contrasted with what is observed to go on in the real-world situation. A relevant system, in that sense, is an entirely abstract idea.

When one or more relevant systems to take forward to the next stage of the approach have been envisaged, each system must be described precisely in words. The technical term given to this written description of the essences of the processes implied by the relevant system is 'root definition'. For example, a study of the role of an inner city community centre came up with the following definition of one relevant system: 'an institution encouraging and helping community action aimed at development of the community's own resources'. What this definition expresses is that the essence of the system is encouraging and helping; these are the fundamental processes implied by this particular system, and the root definition tries to capture and express them in as economical and pithy a way as possible.

Guidance on writing root definitions is available in the form of a checklist referred to by the mnemonic CATWOE. According to this, when writing a root definition you should take into account the aspects given in Table 4.1.

Table 4.1   CATWOE

	Stands for	**Meaning**
C	Customers of the system	In this context, 'customers' means those who are on the receiving end of whatever it is that the system does
A	Actors	The individuals and groups who would actually carry out the activities envisaged in the notional system being defined
T	Transformation process	The process by which inputs to the system are transformed into outputs
W	World view (or Weltanschauung)	The world view(s) that guided and informed the selection of the relevant system being defined
O	Owner(s)	The owners of the notional system being defined
E	Environmental constraints	The constraints imposed by the system's environment that are taken as given

It is not the intention that customers, actors, and so on should all be spelled out in every definition; indeed, a root definition that did so would probably be very cumbersome. The purpose of CATWOE is to force you to ask searching questions about the definition being drafted. It requires you to ask: 'does my definition include C (or A, or T, or W, or O, or E), and if not, should it?' In the case of T, the answer should always be 'yes' but for the others the answer may sometimes be 'no'.

Stage 4: Conceptual models

Stage 4 entails deriving an activity sequence model, called a conceptual model in the approach's terminology. This model contains all the essential activities that the notional system would logically have to perform. It is important to stress that the model is a purely abstract one. It is not a picture of some real-world system nor a system that someone is going to try to build in the future. It is derived solely from the root definition using only the rules of deductive logic. Box 4.4 provides some guidelines for building conceptual models.

BOX 4.4   GUIDELINES FOR BUILDING CONCEPTUAL MODELS

1   The model is a model of a human activity system. Its elements are therefore activities, and will be represented on paper as verbs. The modelling language available to you for constructing the model is therefore all the verbs in the vernacular in which you are working. ... So the first thing to do is to scrutinize the root definition carefully and write down the list of verbs which you think are implied by it. Thus if your root definition described

---

'a system to transform raw material into finished product', that implies – logically – at least two activities: 'obtain raw materials' and 'transform into finished product'.

2   Having got your list of verbs, arrange (or rearrange) them into a logically coherent order. In the above case, for example, the activity 'obtain' would logically have to precede the activity 'transform'.

3   You ought to aim to have a complete account of the system in a smallish number of main activities – somewhere between 6 and 12 is what I would aim at. (For beginners, the guideline is: the fewer the better.)

4   Having arranged your main activities into a logical sequence, next examine each in turn, asking whether it logically implies its own set of subsidiary or back-up activities. If the answer is yes, write these down as verbs, and arrange them in logical order around their front-line activities. Thus, the main activity 'obtain raw materials' might imply secondary activities like 'specify quality and delivery requirements', 'identify suppliers', etc.

5   Your model will now have clusters of primary and secondary activities. Scrutinize the clusters carefully to see whether some have functional points in common. For example, if you have activities involving, say, monitoring and target-setting, then it may be reasonable to group them together as a control sub-system. Other sub-systems may also suggest themselves.

6   In building your model, you are not allowed to introduce any 'real-world' considerations into it. You are required only to deduce what is logically implied by the root definition – and nothing else. Thus you are not allowed, for example, to put the names of real-world departments or entities into the model. ... Ignoring the considerations of the real world is, however, easier said than done, and there is a noticeable tendency for conceptual modelling to slip back into the construction of models of systems known to exist 'out there'.

The requirement that one avoids real-world considerations may seem perverse, but there are two good reasons for it, both of them concerned with keeping the conceptual model as detached from the problem situation as possible. The first is that when one makes the transition from stage 2 to stage 3 [of the approach], one crosses into the abstract world of systems thinking. The point of doing that is to develop – using systems ideas – an alternative view of the problem situation which can be compared with it in due course. This alternative view is embodied in the conceptual model, and if it is to remain genuinely detached from the actual situation, then it should be as untainted as possible by the considerations which apply in that situation.

The second reason for excluding the real world relates to the distinction between what one might call 'whats' and 'hows'. The model is constructed in terms of 'whats' – activities specifying what, logically, must go on in the system you have defined. But the model should not be concerned with, or specific about, how these logically required activities should be carried out; that is a matter for discussion much later in the process, if at all. 'Whats' are general and belong to the world of abstractions. 'Hows' are specific, real-world ways of carrying out 'whats'. Thus a digital computer is a 'how' in relation to the 'what' called 'information processing'.

7   The logical nature of the relationship between root definition and conceptual model means that it makes sense to think about definitions and models as linked pairs. Since all the stages of the approach (including this one) are iterative in any case, what you will probably find yourself doing is going back to revise the definition as you struggle with the construction of the model. There is nothing necessarily wrong with this, so long as you make sure that the model and its definition are a logically defensible pairing, and that you are not simplifying your root definition simply because of the difficulties of constructing a complex conceptual model.

(Source: T306 Block 2, Part 3, pp. 107–8)

An example of a conceptual model of a package delivery system is shown in Figure 4.14.

Stage 5: Comparison of models and real world

In stage 5 the conceptual model is compared with what is perceived to exist in the actual problem situation, so this stage marks a return from the 'abstract world' of systems thinking to the 'real world'. Two kinds of outcome are likely to emerge: first, a reassessment of the problem setting, and in turn perhaps some different ideas for relevant systems; and second, an agenda of possible changes that can be taken forward to stage 6.

There are a number of ways of carrying out the comparison stage. One way is to use the conceptual model to devise a set of questions that are answered with specific reference to the real-world situation. Typical questions might be as follows:

Here is an activity in the model: does it happen in the real world?

If so, who does it, and why?

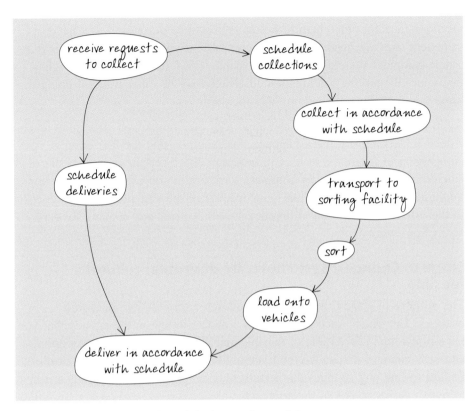

Figure 4.14 A conceptual model of a package delivery system

What is the history of it in the real situation?

Why is it done in the way it is?

Is there a particular reason for doing it in this way?

Another way is to imagine the conceptual model operating and look at something that would happen according to the sequence of activities in the model. Then compare that with how those activities or their closest counterparts would happen, or have happened, in the real situation. For example, if you were looking at a hotel's reservation system, you could compare the conceptual model for receiving, processing and confirming a booking with how a request for a reservation is handled at present.

Another option is to use the rich picture to build a model of the real-world situation that is structured, as closely as the situation will allow, in an analogous way to the conceptual model. If this can be done – and it will work only for situations that are already fairly well structured – you can then make a direct comparison between the two.

However the comparison is carried out, the agenda of possible changes that emerges from it should always be couched in general terms of 'whats' – such as what activities are missing, or problematical, or questionable, or whatever – and never in terms of specific 'hows'. Thus an agenda could include a statement like 'some systematic way of centrally processing reservations across the chain of hotels seems to be necessary', but never a statement like 'we need a new computer system'. The reason for this is that a new computer system is just one possible 'how' for achieving the 'what' that your comparison has highlighted. There may be other ways of doing the same thing, such as outsourcing the operation.

As in each preceding stage of the approach, the comparison stage may reveal a need for iteration. It often happens, for example, that the first thing a comparison highlights is lack of information about certain aspects of the situation, in which case it is necessary to return to stage 2. The comparison might also cause you to rethink your reading of the situation, perhaps even to the extent of identifying different relevant systems and moving forward again from stage 3.

Stage 6: Changes: systematically desirable; culturally feasible

The purpose of stage 6 is to identify and agree feasible and desirable changes. The device used to do this is a debate based on the agenda that you have drawn up. The criteria of feasible and desirable are equally important, and only changes that satisfy both should be considered for implementation. Failure to find any changes that satisfy both criteria therefore signals a need to revisit earlier stages of the approach.

Stage 7: Action to improve the problem situation

Agreed changes are likely to take the following forms:

1   Changes in structures – these may be changes in organisational groupings, departments, reporting structures, lines of command, lines of functional responsibility, or even physical layout.

2   Changes in procedures – alterations in the dynamic elements in the situation (the processes or activities that go on within it). Changes of this type really just amount to different ways of doing things.

3   Changes in policy – in the goals and strategies of the human activity system(s) being investigated.

4   Changes in culture.

This version of SSM gives very little explicit advice on implementation. For the first three types of change listed above, the implementation techniques introduced in Block 3 will be very relevant. These changes are likely to be relatively easy to implement provided that the people involved agree about their desirability and feasibility. Changes in culture, however, are much more problematic. I shall look at culture in Block 5.

5.2   Evaluating the seven-stage model of SSM

The main criticism levelled at the seven-stage model of SSM is lack of comprehensiveness, particularly in the later stages of the analysis and design process. It is certainly undeniable that the approach is strongest in the early stages of problem identification and analysis. One of the merits of SSM, and its emphasis on 'rich picturing' of the situation in the early stages, is that it is less likely than many other approaches to overlook the interactions or interconnections that may give rise to unintended consequences of implementation. Nevertheless, even agreed changes can have unintended consequences and their implementation may lead to new problems and a new problem situation. Therefore, post-implementation review, possibly leading to further iteration, is bound to be important even though it is not part of the approach.

Now read Offprint 8, which describes an application of SSM to bring about improvement.

6   THE SYSTEMS FAILURES METHOD

Situations where there are problems are often seen as being in a very different category from those that have experienced a disaster or are regarded as failing, but there are significant areas of overlap between them. Indeed, it is frequently only the question of scale that allows observers to make a clear distinction. For example, materials defects in one component which result in the recall of a particular make and model of car may share many common features with the gauge corner cracking that led to the Hatfield railway disaster in October 2000 in which an express passenger train came off the tracks at 115 mph, killing four people and injuring 33. However, the former case would be treated as a quality problem and the latter as a failure.

In carrying out improvement exercises, higher-level, over-arching problems with greater consequences are often neglected. For example, once an SPC scheme has been introduced and is seen to work well, investigations into future problems are likely to be confined to individual processes, even in cases where the roots of the problems lie within the control system itself. Because individuals experience different, smaller aspects of the same larger problem a huge amount of evidence may have to pile up before the scheme itself is suspected, and even then the problem solvers may lack the means or motivation to tackle complex, poorly defined problems. The Systems Failures Method offers a way of clarifying, investigating, and dealing with such higher-level problems or problem messes.

At its heart the Systems Failures Method has two key features: conceptualisation and modelling of the problem or failure situation as a system or systems; and comparison of that system, initially with a model of a robust system that is capable of purposeful activity without failure, and subsequently with other models based on typical failures. The activity of conceptualising and modelling systems as a prelude to studying a situation can itself be likened to a transformation process. It begins with users of the method being confronted with a situation that someone has identified as a problem or a failure and ends with their having sufficient knowledge of the situation in systems terms to be able to represent it, or aspects of it, in the appropriate format to allow the comparisons.

6.1   The stages of the Systems Failures Method

The stages of the Systems Failures Method are shown in Figure 4.15. I shall describe each of the stages briefly here. A more detailed explanation of the method and examples of its use can be found in Fortune and Peters (1995; 2005).

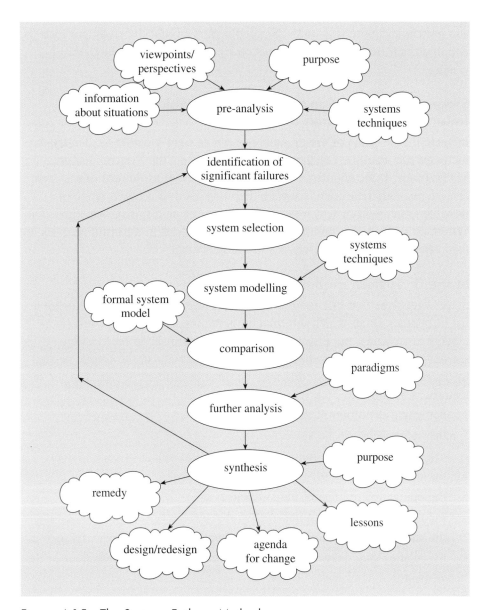

Figure 4.15   The Systems Failures Method

Stage 1: Pre-analysis

The first stage entails deciding which aspects of a situation are being regarded as the failure(s) or potential failure(s), and hence the focus of the application of the approach, and which of the many systems that could be conceptualised within the situation are likely to advance understanding. It is also necessary to define the purpose of the analysis and the viewpoints and perspectives from which it is being carried out, and to gather and organise information about the situation. Techniques that can be used at this stage include: spray diagrams; relationship diagrams; multiple-cause diagrams; rich pictures; and non-diagrammatic methods such as lists, databases, charts and the like. The decision about which techniques to use in the pre-analysis of any particular failure rests with the people carrying out the study but there is one important rule: *the situation must not yet be represented in terms of systems.*

Stage 2: Identification of significant problems and/or failure(s) and selection of system(s)

In stage 2 all aspects of the pre-analysis are brought together to identify the focus for the analysis. For the purposes of analysis, the Systems Failures Method regards the significant problems or failure(s) as outputs of a system or systems. Different boundaries are therefore tried to create a range of possible systems from which the failure(s) can be said to have emerged. The system or systems that appear the most likely to lead to a fruitful analysis are then selected to be carried forward to the next stage.

Stage 3: System modelling

Stage 3 involves first clarifying the nature of the system(s) selected using a list of features about which information is required. Then diagrammatic models of structure and process are built as a precursor to representing various aspects of the system(s) and system behaviour in the format needed for stage 4 of the approach; this may be done at different levels. The list of features is as follows:

1   name and definition of the system(s)
2   the components of the system(s)
3   the components and relationships in the environment of each system
4   the wider system
5   the inputs and outputs of the system(s)
6   the main system(s) variables
7   the structural relationships between components
8   the relationships between the variables that describe the behaviour of the system(s).

The diagrammatic models of structure and process are built using input–output diagrams, systems maps and influence diagrams. For example, Figure 4.16 shows a systems map of the science education in primary schools system as perceived immediately before the introduction of the National Curriculum. The map was part of a study prompted by quality problems identified in reports by schools inspectors (HMIs) throughout the 1980s (DES, 1983; 1985; 1990).

Stage 4: Comparison

In stage 4 a comparison is made between the system that has been selected and a model that unites most system concepts. This model, called the Formal System Model (FSM), was adapted from Checkland (1981), who in turn drew on the ideas of Churchman, particularly his concept of a teleological

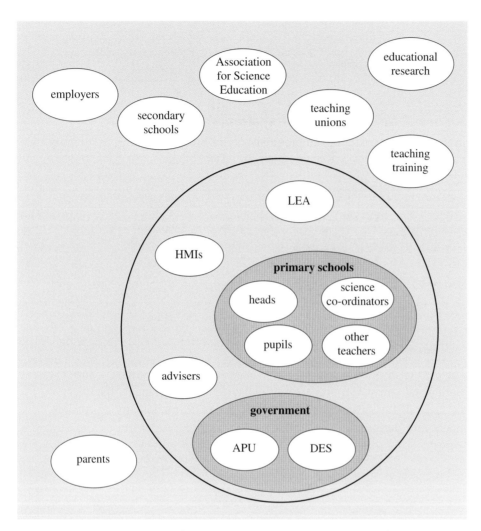

Figure 4.16   Systems map of the science education in primary schools system
(Source: Fortune et al., 1993, p. 363)

system (Churchman, 1971, pp. 42–78), and Jenkins (1969). The term 'formal' is used in the title to indicate that the model is a form or structural framework into which something can be fitted. The bare bones of the framework are shown in Figure 4.17. These are: a system (the Formal System), a wider system, and an environment.

The wider system represents the next hierarchical level upwards from the system. It affects the system in a number of ways:

● It defines the purpose of the system and sets objectives for it.
● It influences the decision makers within the system and monitors the performance of the system as a whole.
● It provides the resources that the system needs in order to function.

The environment disturbs the system directly and also disturbs it indirectly through the wider system. Similarly, the system attempts to influence the

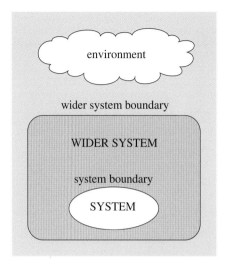

Figure 4.17   System, wider system and environment (Source: Fortune and Peters, 2005, p. 117)

environment directly and via its wider system. These relationships are summarised in Figure 4.18.

The Formal System consists of a decision-making subsystem, a performance-monitoring subsystem and a set of subsystems and elements that carry out the tasks of the system and thus effect its transformations by converting inputs into outputs.

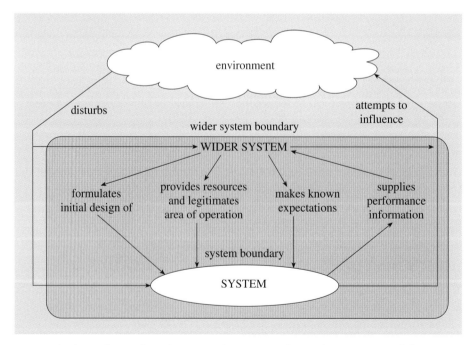

Figure 4.18   Relationships between the system, the wider system and the environment (Source: Fortune and Peters, 2005, p. 118)

The decision-making subsystem manages the system:

- It is responsible for decisions about how the purposes of the system are to be achieved, for example which transformations are to be carried out and by what means, and for providing the resources to enable this to happen.
- It makes known its expectations to the subsystems and components that carry out the system's transformations and to the performance-monitoring subsystem.

It is therefore the decision-making subsystem that allows the system to exhibit choice, and thus behave as a purposeful system.

The job of the performance-monitoring subsystem is to observe the transformation processes and report deviations from the expectations to the decision-making subsystem so that it can initiate corrective action where necessary.

These three subsystems and the relationships between them are shown in Figure 4.19. An equivalent relationship to that between a wider system and a system can be said to exist between the Formal System and its subsystems.

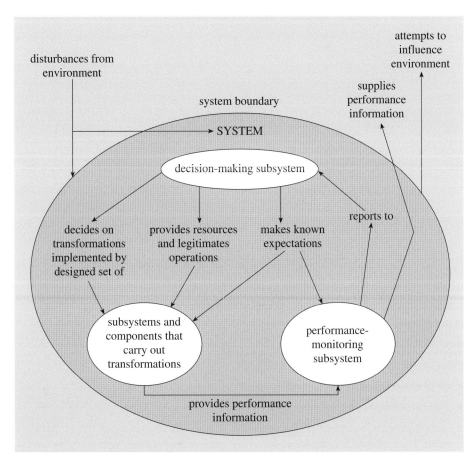

Figure 4.19   Relationships between subsystems (Source: Fortune and Peters, 2005, p. 119)

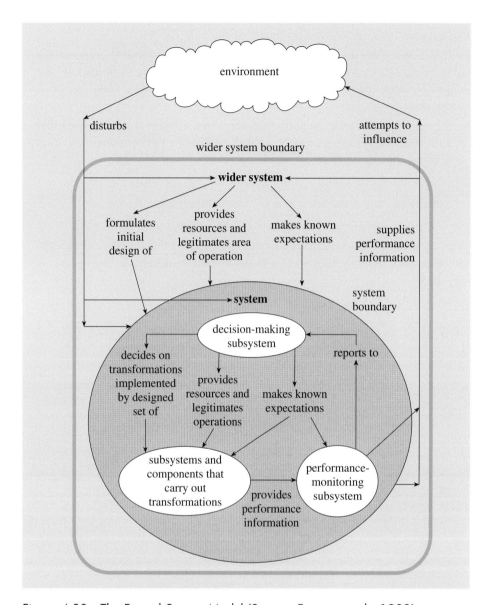

Figure 4.20   The Formal System Model (Source: Fortune et al., 1993)

By definition, a Formal System has requirements of its subsystems, but each subsystem still has a certain amount of autonomy in deciding how those expectations are met.

The comprehensive representation of the FSM is shown in Figure 4.20. Its features can be summarised as follows:

- a decision-making subsystem
- a performance-monitoring subsystem
- a set of subsystems and elements that carry out the tasks of the system and thus effect its transformations by converting inputs into outputs
- a degree of connectivity between the components
- an environment with which the system interacts

- boundaries separating the system from its wider system and the wider system from the environment
- resources
- a continuous purpose or mission that gives rise to the expectations
- some guarantee of continuity.

Insights gained from comparison between the FSM and the system conceptualised from the situation will be found within subsystems and in the links between subsystems, or will be associated with one or more of the following interfaces between boundaries:

- environment–wider system
- environment–system
- wider system–system
- interfaces between subsystems.

Because of the hierarchical nature of the FSM each of its subsystems can also be perceived as a Formal System with its own decision-making, performance-monitoring and transformation-effecting components, and each one of those could then, in turn, also be regarded as a Formal System, and so on.

Wide experience (see Peters and Fortune, 1992) of comparing systems representations of failure situations with the Formal System Model has revealed recurring themes. The following are typical points of difference:

1   deficiencies in the apparent organisational structure of the system, such as a lack of a performance-measuring subsystem or a control/decision-making subsystem
2   no clear statements of purpose supplied in a comprehensible form to the system from the wider system
3   deficiencies in the performance of one or more subsystems, for example the performance-measuring subsystem may not have carried out its task adequately
4   lack of an effective means of communication between the various subsystems
5   inadequate design of one or more subsystems
6   not enough consideration to the influence of the environment, and insufficient resources to cope with those environmental disturbances that were foreseen
7   an imbalance between the resources applied to the basic transformation processes and those allocated to the related monitoring and control processes, perhaps leading at one extreme to quality problems and at the other to cost or output quantity problems.

In order to undertake the process of comparison it is first necessary to represent the system that has been conceptualised in the same format as the FSM. In Figure 4.21 this has been done for the science education in primary schools system.

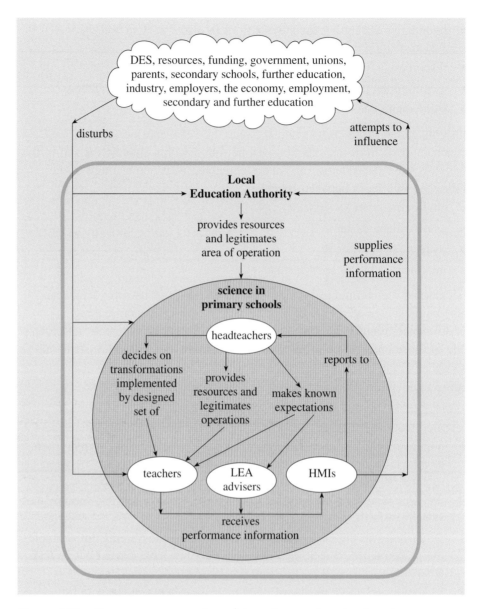

Figure 4.21   Systems representation of the science education in primary schools system as a basis for comparison with the Formal System Model (Source: Fortune et al., 1993, p. 364)

Comparing a representation such as that in Figure 4.21 with the FSM yields a set of discrepancies that form the output of this stage. In this example the elements and/or subsystems of the system appear to be disparate and to receive different information from the decision-making subsystem. The second discrepancy concerns the decision-making subsystem itself; it does not make its expectations known to the performance-monitoring subsystem. A third point to emerge concerns the performance-monitoring subsystem; it supplies performance information not only to the wider system, but also to the environment, and thus represents the system's main attempt at influence.

Finally, the wider system has not formulated the initial design of the system (this has been done by a subsystem, the advisers) and has not made known its expectations (this has been left to the decision-making subsystem).

In order to generate sufficient understanding it is often necessary to make further comparisons against the FSM at different hierarchical levels. These can be upwards and/or downwards. When moving up, the system which has just been considered will next be viewed as a subsystem of a higher-level system. When moving down a level, what was regarded as a subsystem will be perceived as the system. Judgement about whether a change in level is appropriate, and if so whether it should be up to a wider system or down to what was a subsystem, will need to take account of the nature of the system(s) as conceptualised earlier in the approach and the results of the first comparison.

Stage 5: Further analysis

Further investigation of the discrepancies between the systems representation of the situation and the Formal System Model takes place using other systems-related models concerned with control and communication.

Control is most obviously relevant to the workings of the performance-monitoring subsystem. Performance monitoring generates the information required in order to exercise control. Control can then be used to minimise, as far as possible, the deviations of the outputs of the system from the values that will enable the expectations to be met. Clearly, there is a strong communication dimension to control, but in addition the following requirements for communication can be highlighted:

1   communication between the system and its environment
2   the flow of information from the wider system, via the system, to the subsystems, and vice versa
3   numerous communication links within the system and the subsystems.

Stage 6: Synthesis

Figure 4.15 showed the main iteration loop within the approach, but it could have included many more loops. At every stage, iteration may be required, both within stages and between them. For instance, gaps may appear in the information gathered during the investigation or it may become clear that choices, perhaps relatively minor ones such as placing a component in the environment rather than in the system at the pre-analysis stage, were inappropriate, thus making it necessary to revisit a particular stage in the approach. Because multiple viewpoints and perspectives are required, it may even be necessary to make a number of passes through the entire approach. In the extreme case a situation that was designed to match the model precisely, and should thus be capable of operating without failure, could still be judged by some observers to be a failure because it relied on a high degree of coercion or encouraged self-exploitation or breaking of the laws governing data protection, for example. The failure would not emerge until these perspectives were considered.

When the analysis starts to look complete it becomes necessary to draw the threads back together. In order to maintain the systemic nature of the investigation the best way to begin the synthesis is to return once again to the Formal System Model and use the findings from all the iterations to remodel the system at the various key levels. Once that has been done the format of the remainder of the synthesis may well vary according to how the results of the study will be used. As a first step it may be helpful to prepare another version of the set of FSMs which emphasise the salient features. What is 'salient' is always a difficult judgement, with the answer depending on the eye of the beholder, but the starting point for determining it will be the viewpoint, perspectives and purpose that informed the pre-analysis.

Facilitating the process of understanding failure or potential failure, learning from that understanding and taking action are paramount and so the synthesis has to be developed and presented to others in a way that supports that process. The findings may suggest that remedy is possible; this is likely if the consequences of the failure were systemic but its causes were simple. Alternatively, the findings may indicate that redesign is necessary; this is likely if the causes and the consequences were both systemic. If redesign is needed, the findings can be fed into a process for generating an agenda for change.

6.2   Evaluating the Systems Failures Method

The criticisms that have been levelled against this approach are more philosophical than practical. They stem from its origins in General Systems Theory and are that it is positivist (Mitev, 2000), underplays the importance of power in organisations (Mansell, 1993) and 'does not have a well worked out theory of subjectivity' (Mansell, 1996, p. 503).

7   THE SIX SIGMA METHOD FOR PROCESS IMPROVEMENT

As you saw in Section 11 of Block 3, the standard measure of process capability in SPC charting procedures is based on the ability to perform consistently within ± 3 standard deviations from a correctly centred mean. At this level of performance a process would be 'in control' 99.73 per cent of the time and hence would be operating beyond the control limits for 0.27 per cent of the time. The basis of Six Sigma is that a process needs to exhibit less variation than this to the point where it is capable of achieving a performance that consistently falls within ±6 standard deviations from the mean. Performance at this level would give the expectation of no more than 3.4 defects per million opportunities. Table 4.2 shows the implications of the difference between levels of performance that are based on a 99% level of quality attainment versus a Six Sigma level of 99.9997%.

Table 4.2   What you would get from a 99% level of quality and the contrasting Six Sigma performance

	99% performance	**Six Sigma performance**
For every 300 000 letters delivered	3000 misdeliveries	1 misdelivery
Out of every 500 000 computer restarts	4100 crashes	<2 crashes
For 500 years of month-end closings	60 months would not balance	0.018 months would not balance
For every week of TV broadcasting (per channel)	1.68 hours of dead air	1.8 seconds of dead air

(Source: Pande et al., 2000, p. 12)

The Six Sigma approach is based on this concept of 6 standard deviations from the mean. It was originated by the US company Motorola and is a registered trademark of the company. Late in the 1970s senior management at Motorola became aware that the quality of the products and services they provided was lagging behind that achieved by their competitors. In 1980 the company appointed a corporate quality officer and the following year established a Training and Education Centre with the aim of improving quality by ten times in five years. At about the same time Bill Smith, a senior engineer and scientist with the company, was concerned at increasing complaints from the sales force about warranty claims and wanted to develop a standard way of measuring quality that would enable different processes to be compared. Bob Garvin, Motorola's chief executive, took up Smith's ideas and lent his weight to them. In 1985 the Communications Products division of the company began to measure total defects per unit, first for manufactured products in July and then for sales orders in November. In 1987 the corporation adopted the goal of achieving a Six Sigma level of performance by 1992.

Mention of 'total defects per unit' could imply that Six Sigma is really suited only to manufacturing processes or those such as order processing and invoicing that share many of the characteristics of manufacturing processes. However, as the list in Box 4.5, which is taken from the health care domain shows, this is not the case.

BOX 4.5   LIST OF SAMPLE SIX SIGMA PROJECTS IN RADIOLOGY

Reducing time between report dictation and report signature

Reducing patient wait time from arrival in radiology to time of exam

Reducing time between patient dismissal and dictation completion

Reducing patient wait time for radiology registration process

Enhancing radiology scheduling process

Reducing time from radiologist signature to report distribution

Increased efficiency in the MRI ordering process

Optimizing the content quality and delivery of pre-exam patient education

Reducing time for dismissal of radiology patients

Enhancing film jacket retrieval process

Decreasing MRI report turnaround time

Improving general radiology staff scheduling

Increasing efficiency of ultrasound exam scheduling and reducing overtime

Utilizing special procedures inventory more efficiently

Augmenting radiology exam scheduling and preregistration process

Reducing CT order to taken time

Decreasing IVP exam time

Improving utilization of nuclear medicine radio-pharmaceuticals

(Source: Snee and Hoerl, 2005, p. 76)

This section will look specifically at the method for process improvement that is part of the Six Sigma approach. There are several versions of the Six Sigma method. Some companies draw a distinction between Six Sigma as a concept and 6S as a method of effecting change. The phases of the 6S method are set out in Figure 4.22. However, in this discussion I shall use the original Motorola methodology. It is known by the mnemonic DMAIC (pronounced de-may-ic), which stands for define–measure–analyse–improve–control.

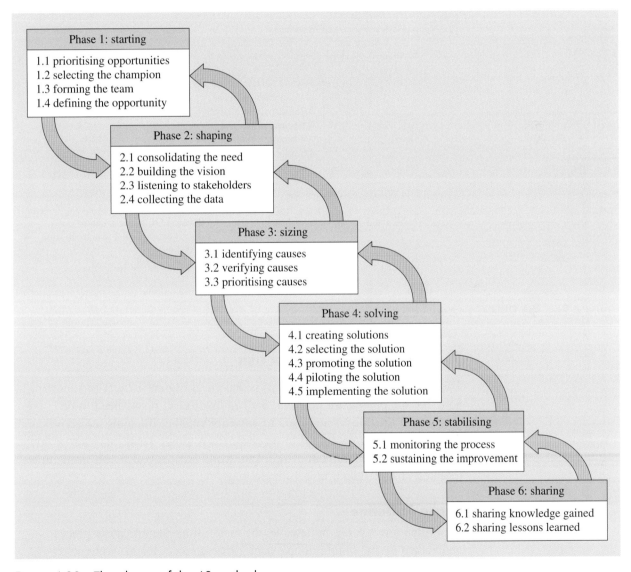

Figure 4.22   The phases of the 6S method

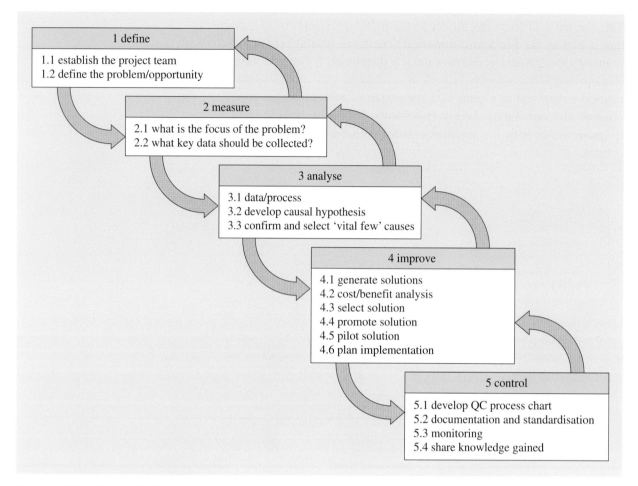

Figure 4.23   The DMAIC process

7.1   The DMAIC process

The DMAIC process (or roadmap) is shown in Figure 4.23. It is important to note that the phases are iterative; the results of each phase need to be assessed as they occur and used to consider whether the aims, objectives and direction of the project need to be changed. A single result may throw doubt on the value or correctness of previous work and suggest that it needs to be repeated or undertaken in a different way.

Phase 1: Define

Six Sigma projects are initiated as the result of an identification process undertaken by the organisation's senior management team. A project champion is put in place at this stage and plays a very important part. In Six Sigma's own special terminology the champion is usually a 'black belt' and thus able to act as instructor, mentor, and expert reference to the project team leader ('green belt').

The champion's job is actively to promote the project and to direct the work of the project team. This involves taking every opportunity to support the project, resolving conflicts and removing any barriers that may arise. The project champion should:

- be prepared to take brave, innovative decisions that challenge the way that operations are currently undertaken;
- provide the project team with a view that is informed by an organisation-wide knowledge of processes, people and other projects that are being worked on;
- ensure that the changes being proposed are not in conflict with other projects;
- play a leading part in the appointment of the project team leader and, to a lesser extent, in the recruitment of team members;
- identify and liaise with other people in the organisation who are likely to influence the success of the project;
- ensure that she or he has sufficient time to devote to the project and the project team and especially to monitor and review progress;
- build support for the change within the senior management team;
- stop the project if changed circumstances or the uncertainty of real benefit make it necessary to do so;
- energise the team, encouraging the members to do good work.

The aim of the first phase of the Six Sigma methodology is to get the project off to a good start by identifying the team that will undertake the work and establishing the way its members will work together, and developing an initial definition of, and plan for, the project. The composition of the team will vary according to the nature of the project but it is likely to have Six Sigma experts including a green-belt project team leader, process experts, data gatherers, someone to represent users and someone to represent customers.

The outputs of the definition phase are:

- an established, enthusiastic project team
- a team charter
- a project charter
- a project plan.

The appointment of the project team leader is perhaps the single most important contribution to the success of the Six Sigma project. The project team leader's job is to manage the day-to-day effort on the project and to initiate, facilitate and co-ordinate the work of team members. The project requires that the team members:

- have demonstrated an enthusiasm and aptitude for improvement activities
- be respected by their peers

- can make sufficient time from their other duties to be able to contribute to the project
- be familiar with, or prepared to learn, analytical and statistical techniques
- be motivated to get involved with the project.

Once the team members have been appointed they can begin work, and the first thing to be done is to establish and agree the team's ground rules. These are usually incorporated into a team contract. The aims of this document are to make explicit the way that the team will operate and to reinforce the commitment of its members to the project.

Once the way in which the team will work together has been settled, the next task is to make an initial definition of the job to be done. This takes the form of the project charter, which clarifies the problem or improvement opportunity that will be addressed. It also provides a reference document to which the team can refer (and revise if necessary) over the course of the project. The reference document, and the process that produces it, are important first steps in the transfer of project ownership from the project champion to the team.

The team may have been given what is sometimes known as an 'initial scoping document', which sets out a brief description of the project and its background, and a definition of the problem or opportunity that the project will address, explaining how it fits with the organisation's business strategy or stating the business case, and outlining why it is important to undertake the project at that time. The project charter is an expansion of these elements and contains as a minimum:

- A brief explanation of the project in the form of a title. This should be long enough to encapsulate and communicate the essential features of the project in a clear and unambiguous way, but should also be as concise as possible.
- A statement of the problem or opportunity that the project will address. (Is/is not analysis can help here.)
- The project's aims and objectives.
- A statement of project outcomes.
- The measures of success of the project.
- The stakeholders in the project. It is unnecessary at this stage to undertake a full analysis of interests but it is important to identify potential stakeholders and their interest in the project.
- An identification of the customers and business impact of the project. This section should contain judgements about the potential impacts of the project. At this stage it is to be expected that the information will be in the form of estimates, but they should be sufficiently credible for the project champion and any other sponsors of the project to either commit to the project as defined or to suggest revisions to the project charter that will increase the benefits. A customer/business impact analysis table, as illustrated in Figure 4.24 can be developed.

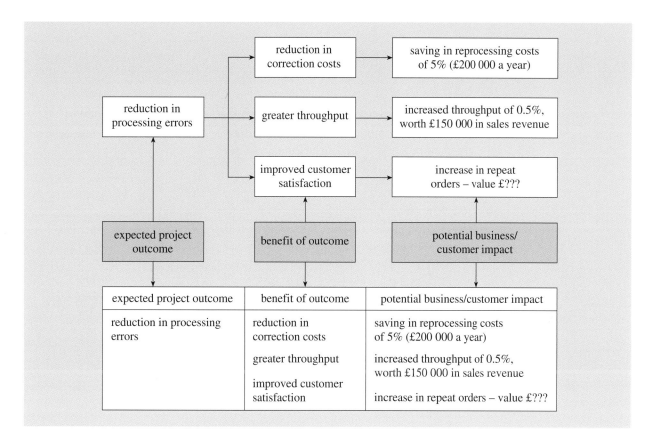

Figure 4.24   Customer/business impact analysis

The next output of the definition phase is a project plan. This can be based on the phases and steps of the Six Sigma methodology and the objectives that have been identified as part of the project charter. It should identify:

- milestones, with target dates for each
- other key target dates for the steps
- activities in each step, the team member responsible and estimated duration
- risks associated with activities, and contingencies
- external resources needed.

One way of thinking about the project charter and plan is to use the five Ws and H technique, which you met in Block 3. The way that this applies to the definition phase is shown in Figure 4.25.

Another analysis that is often undertaken during the definition phase involves the construction of a SIPOC diagram – another technique that was described in Block 3. Figure 4.26 is an example of a SIPOC diagram relating to the process of making photocopies.

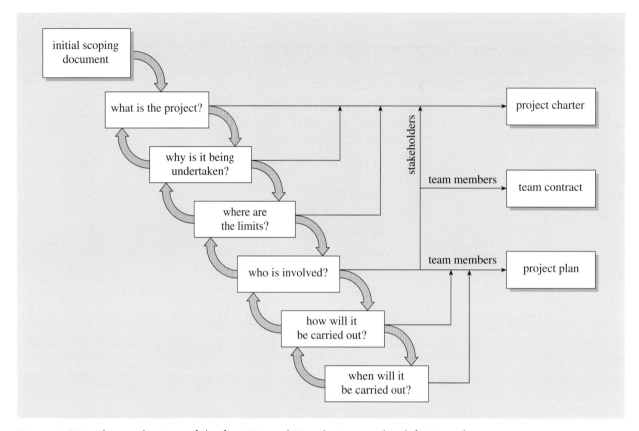

Figure 4.25   The application of the five Ws and H technique to the definition phase

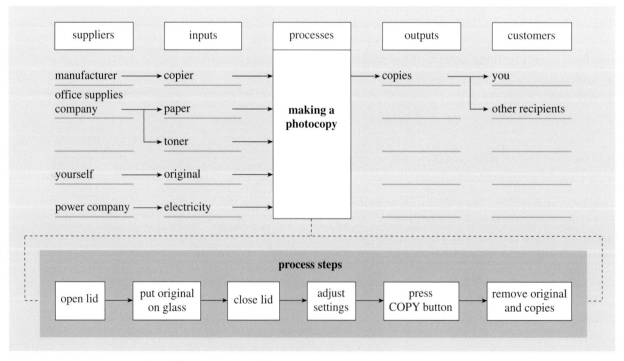

Figure 4.26   An example of the SIPOC model

ACTIVITY 4.5 ....

Construct an input–output diagram for the definition phase. ●

Phase 2: Measure

The aims of the measuring phase of the DMAIC methodology are to elaborate and refine the initial view of the problem or opportunity that was developed during the definition phase and to establish a 'baseline measure' of the current performance of the process. This involves:

● developing a full understanding of customer requirements

● defining how the process operates in practice

● listening to the views of stakeholders in the process

● defining and collecting data.

The aims of the measuring phase can be viewed as capturing the voices of the customer, the process and the people, as illustrated in Figure 4.27.

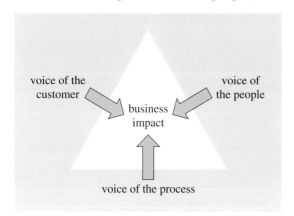

Figure 4.27   The three aspects of the measuring phase

Many people are familiar with the term 'voice of the customer', used to describe stated and unstated customer needs or requirements. 'Voice of the process' is a less widely recognised term but it is of equal importance here. The reason for its significance is that it is the operations process that plays a large part in creating customer satisfaction. One of the outputs of this phase is a set of process flow charts that set out the detail of the physical and logical arrangement of the process. The various forms of process flow charting have been discussed in Block 3. The level of detail required is a matter of judgement for the project team but the general applicability of the main forms of flow chart is shown in Figure 4.28.

The third side of the triangle shown in Figure 4.27 involves capturing the 'voice of the people'. An important part of the work undertaken during this phase of the methodology is listening to the stakeholders in the process. This performs two functions: first, it will give the members of the project team insights into the performance of the system; second, it begins the important

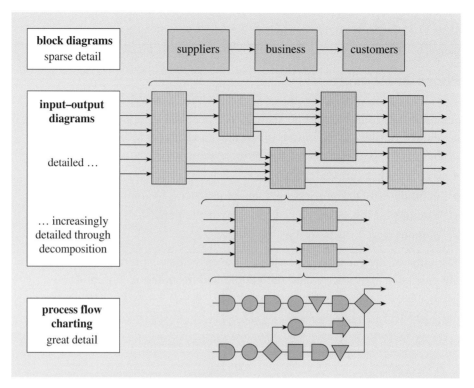

Figure 4.28   Levels of process analysis

task of increasing the commitment of stakeholders to the changes that will take place. Thus, the work of this phase involves soft data, such as views and opinions, but this should be complemented by the collection of hard data. This will provide a detailed picture of how the process operates, pinpoint its weaknesses, establish the performance of the process that can be used to measure the value of improvements, and give a fuller picture of the relationship between causes and effects. The importance, within the Six Sigma ethos, of being able to establish *and demonstrate convincingly* a cause–effect relationship cannot be over-emphasised.

ACTIVITY 4.6 ....

Construct an input–output diagram for the measuring phase. ●

Once the data and other information have been collected they can be analysed, and this takes place in the next phase of the DMAIC methodology.

Phase 3: Analyse

Once clear information on what customers want has been developed, the way in which the process operates has been diagrammed, the opinions and views of the stakeholders gathered and data collected on those important aspects of the process, it is time to bring these aspects of the problem or opportunity together and to establish relationships between them.

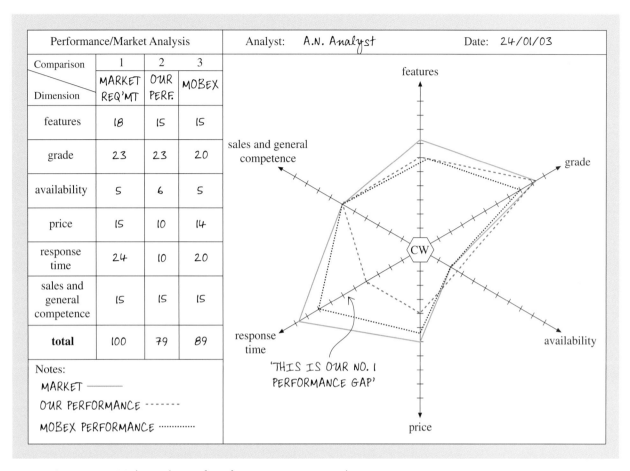

Figure 4.29   EuroMob analysis of performance versus market requirements

As an example I shall look at EuroMob, a firm specialising in the manufacture and supply of mobile hospitals for use in war zones, and for famine and disaster relief. These hospitals could be configured in a variety of ways but typically consisted of a number of mobile vehicles used as operating theatres, intensive care units and similar functions, and tents used for recovery wards and for administrative and staff purposes. EuroMob was responsible for the design, manufacture and supply of the hospital as a whole, but subcontracted the manufacture of the tents to a specialist producer of this type of equipment. It was losing orders to rival companies and in particular to its main competitor, Mobex, so a study of performance versus market requirements was carried out. This resulted in the analysis shown in Figure 4.29. It revealed that 'response time' was the major competitive issue for the company, especially as the analysis suggested that the company's performance in this aspect was falling behind that of Mobex. Customers placed orders as the result of a crisis or urgent need and, although they did not expect instant availability because each configuration was unique, they did want a fast response.

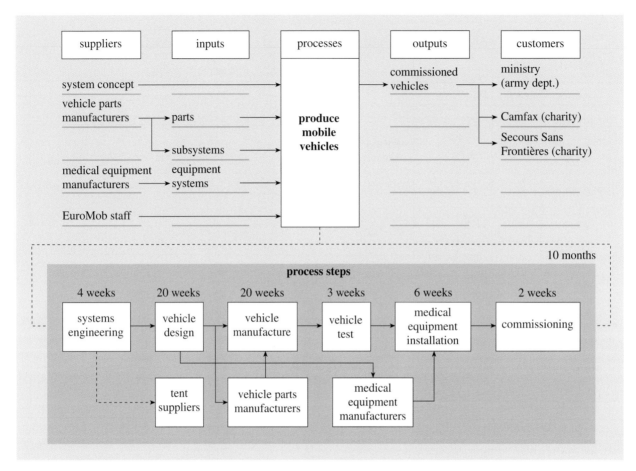

Figure 4.30   The EuroMob SIPOC diagram

In the light of this analysis EuroMob's senior management team initiated a Six Sigma project that, put simply, had the aim of reducing 'concept to customer time'. The SIPOC diagram developed for the high-level relationships associated with mobile vehicles is shown in Figure 4.30. It indicated that the two longest processes were design and manufacture, but after discussions with the managers concerned it was decided to focus on the design function.

The design process involved developing a concept for the vehicle, including its medical and other equipment, and then testing that concept using simulation and finite element analysis. The software used for these tasks was costly, difficult to learn, and would run only on a high-specification personal computer. Because of the expense of these workstations, EuroMob had bought only one. Two designers had been trained to use the software but one of them was no longer available to do much, if any, of the simulation and analysis (S&A) work. It is not surprising, therefore, that the simulation and analysis workstation was a bottleneck in the design process. Senior designers were continually having to chase work through the workstation and to rearrange work schedules.

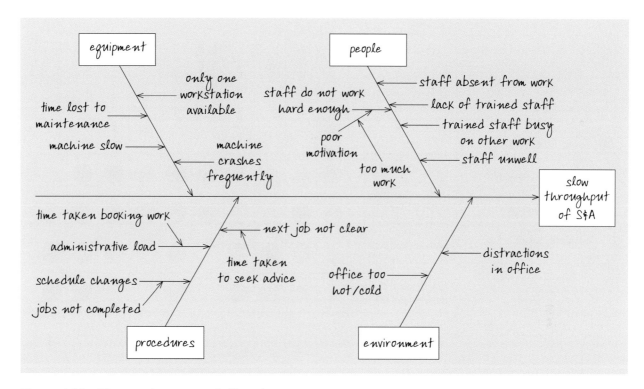

Figure 4.31   The team's cause-and-effect diagram

One strange feature of this situation was that everybody in the design office was conscious that simulation and analysis were a bottleneck – this became clear during the activity of gathering the 'voice of the people' – but other than grumbling about it occasionally, nothing had been done. Managers and senior designers were not aware of the effect of the delays caused by the bottleneck.

Before the Six Sigma team members started collecting data on the dynamics of the bottleneck they undertook a cause-and-effect analysis of the problem of 'slow simulation and analysis throughput', and produced the diagram shown in Figure 4.31. This diagram was used as the basis of a Y2X analysis which was undertaken by the Six Sigma team and a cross-section of design office staff. The outcome of this analysis is shown in Figure 4.32.

The process that the EuroMob Six Sigma team went through illustrates an important characteristic of the Six Sigma methodology: the requirement to establish and demonstrate a relationship between the detailed way in which operations are carried out, customer requirements and the strategy of the organisation. This set of relationships is illustrated in Figure 4.33, and the whole process is one of progressively focusing on the 'vital few' changes that can have the greatest impact on customer requirements.

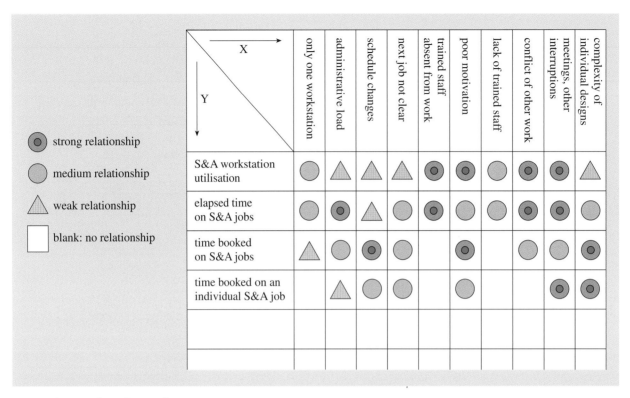

Figure 4.32   The Y2X analysis

A way of focusing attention on what needs to be changed in an operations system is to identify where 'waste' occurs. The topic of waste, which was introduced briefly in Block 3, is synonymous with the term 'non-value-added' (NVA). The Japanese industrialist Taiichi Ohno classified the waste that occurs in operations systems into seven categories (Imai, 1997, p. 75):

- overproduction
- inventory
- repairs/rejects
- motion
- processing
- waiting
- transport.

To these categories (sometimes called the 'old wastes') have been added a further five 'new waste' categories:

- human potential
- excess energy
- pollution
- space
- complexity.

An explanation of various categories of waste is given in Box 4.6.

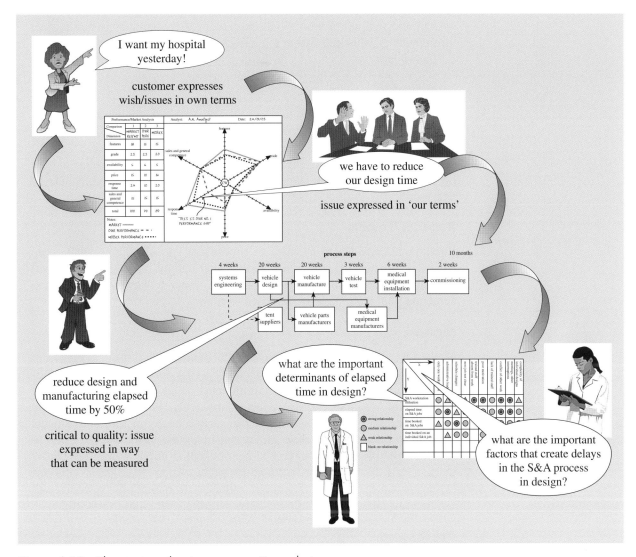

Figure 4.33 The strategy/customer–operations chain

BOX 4.6 THE FORMS OF WASTE THAT MAY OCCUR IN AN OPERATIONS SYSTEM

Overproduction

This may occur for many reasons but most are due to uncertainty. For example, a project team leader will produce more than the schedule requires because he or she is concerned about the reliability of downstream processes. The extra work is produced just to be 'on the safe side'. The measurement system may also contribute to overproduction. If a manager knows that the first figure that the managing director looks at on Monday morning is the throughput of the system and that sharp words will be spoken if the amount is down, then throughput is what the managing director will get *whether the output is needed or not.*

Inventory

Raw materials, parts, work-in-progress and finished goods are all forms of waste which add to cost until the money for them is received. Traditional accounting practice regards inventory as an asset to be entered on the balance sheet of the organisation but as far as the operations system is concerned it is a liability (in the everyday use of that word).

Repairs/rejects

Repairs and rejects are a waste of productive capacity since the system has made an item or rendered a service that cannot be sold and that will incorporate the value of its inputs. A second waste occurs when someone puts them right or disposes of them.

Motion

Many of the actions and movements made by those involved in operations do not add value. An example where the work of an operator involved processing electronic components is as follows. Part of her responsibility was to heat-treat the components, but the oven that she needed was 300 yards away. She had to carry a heavy rack of components to and from the oven six times a day but made the journey more often to check on the process. An oven had been installed in the area in which she worked but it had never worked properly. Apart from such obvious examples of the waste of motion, almost all operations systems will provide instances that could have been avoided with a little foresight.

Processing

Processes that initially may seem to be adding value to the product or service can, on closer examination, be revealed as wholly or partially unnecessary and are a consequence of poor technology or methods. Examples of this form of waste include de-burring pressed metal components, machining castings, and removing drips and spills from the edges of plates before they are taken to the table in a restaurant.

Waiting

The time spent by an operator waiting for a process to complete its cycle is a waste.

Transport

This is one of the most common and pervasive forms of waste. Even a cursory inspection of an operations system will reveal many examples of transport that could be avoided.

Some time ago some of the body panels for the much-loved MGB sports car were pressed at a plant in Castle Bromwich in Birmingham in the English Midlands. They were then taken by road to another plant at Swindon about 70 miles away. There they met up with other panels that had been produced locally and were assembled into what is known as a body-in-white. These bodies were then sprayed with a wax preservative (a waste of processing) and transported the 40 miles to the Pressed Steel Fisher plant at Cowley on the outskirts of Oxford. The first operation carried out at Cowley was to remove the wax preservative. The bodies were then painted and partially trimmed, that is, items such as the carpets, seats, windows, interior panelling, electrical items and the dashboard were fitted. The bodies were then taken to the MG plant at Abingdon, a distance of 15 miles, to have the power train and running gear fitted, the trim finalised and to be road tested.

By this time the panels produced at Castle Bromwich had travelled a distance of 125 non-value-adding miles. This might be regarded as being merely amusing were it not for the fact that during the early 1970s an adverse movement in the sterling/dollar exchange rate made the MGB much more expensive in the USA, one of the major markets for the car. The resulting drop in sales reduced throughput and it became uneconomic to continue production of the MGB. The Abingdon plant was closed with the loss of 1200 jobs.

Human potential

The waste of the potential of the human beings in an operations system can take various forms, but two are particularly insidious. The first is when operators and others are supposed only to do their job and are not asked or expected to contribute to improvement activities. An old saying is that 'with every pair of hands comes a free brain', and Six Sigma project teams should take every opportunity to make use of the expertise that is often unrecognised. The second waste of human potential occurs when people are required to do jobs that are more effectively carried out by machines. This is often a contentious issue and the history of objections to the replacement of human physical and mental labour by mechanical and electrical power and software goes back to the Industrial Revolution and beyond.

Pollution

Since the work of the Club of Rome in the 1960s (Meadows et al., 1974) attention, in the developed world at least, has been focused on the need to reduce the various forms of pollution that result from all forms of economic and social activity. The result has been a series of measures that encourage (or coerce, depending on your point of view) business to reduce the amount of physical and gaseous waste that it produces.

Space

Space costs money and encourages other forms of waste such as unnecessary internal transport and motion.

Complexity

The final category of waste is the unneeded complexity that is designed into products and systems by engineers and that contributes to many of the other categories.

As Figure 4.34 and the EuroMob example suggest, the analysis phase starts with the problem, a view of the process and data that have been carried forward from the definition phase. Cause-and-effect and Y2X diagrams may also have been produced. These aspects of the problem will point towards likely causes of the effect that is to be eliminated in the case of a problem or enhanced in the case of an opportunity. Once the likely causes have been identified, a hypothesis can be developed to link cause and effect. Hypotheses are not formulated as questions. They can take either a 'That ...' form or an 'If ... then ...' form. Figure 4.35 shows a problem reformulated as both a 'That ...' and an 'If ... then' hypothesis. Both the forms of hypothesis shown provide a clearer basis for the design of the data-gathering exercise and the analysis of the results.

Many attempts at operations system improvement are expedient and temporary in character. As a consequence the problem recurs time and time again or the opportunity is not grasped fully. Effort and time are wasted and the result is frustration and then loss of motivation. In contrast, Six Sigma aims to 'do it right: do it once.'

The examples shown in Figure 4.35 are what might be termed 'business hypotheses' and should be distinguished from statistical hypotheses. It is probable that more data will have to be collected during the analysis phase as the Six Sigma team's understanding of the problem, cause, and effect deepen. But data are not the same as information and it is important to analyse the data to wring meaning and implication from them.

The first step in analysing data is often to use descriptive statistics like those you studied in Block 2, for example:

- calculation of the mean, median and mode of the set of data
- allocation of the data to a class and construction of simple charts such as histograms, pie charts and Pareto diagrams
- calculation of the standard deviation of the members of the data set
- construction of graphs illustrating time series
- scatter diagrams.

Time and cost constraints will usually ensure that the data collected within the context of a Six Sigma project will be a sample selected from a wider population. It is important to ensure that the sample collected is representative of the population from which it is drawn. For example, suppose that in the case of EuroMob the data on simulation and analysis of workstation utilisation were collected in the week before the Christmas

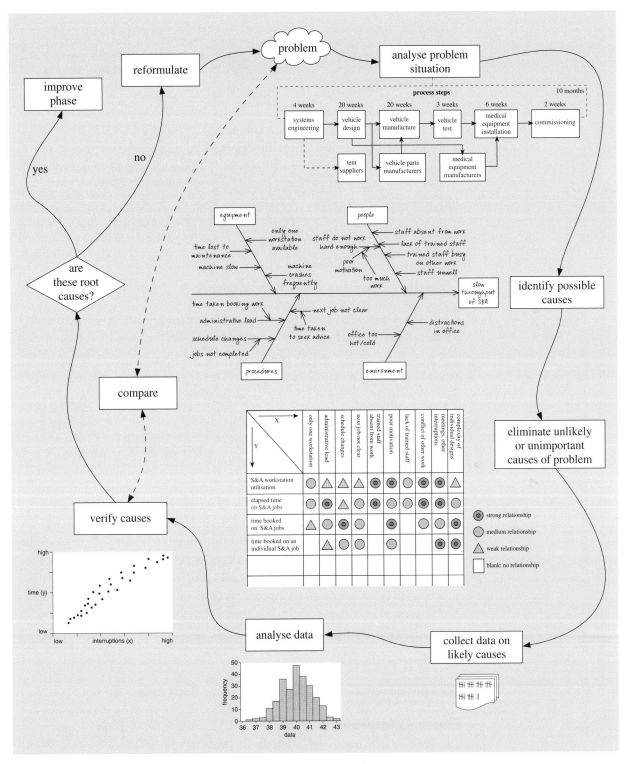

Figure 4.34   The activities undertaken during the DMAIC analysis phase

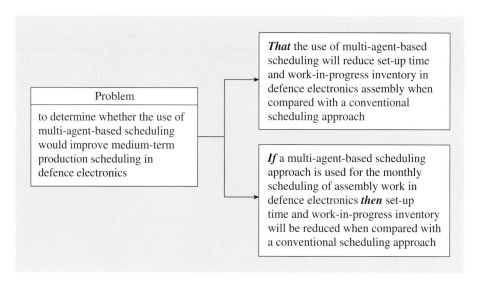

Figure 4.35   Examples of the 'That ...' and the 'If ... then' forms of hypothesis

break. They are unlikely to be representative of patterns of work that would occur in more normal weeks.

The measuring phase of DMAIC is what Ryan (1989, p. 65), following Deming (1953; 1975), has called an *enumerative study.* It is conducted for the 'purpose of determining the current state of affairs' or 'how many'. In contrast, the *analytical study,* carried out during the analysis phase, is focused on determining the causes (Xs) of certain relevant outcomes (Ys) with a view to removing or reducing their effect or inducing an effect that is desired. An analytical study answers the question 'why?'. One of the great strengths claimed for the DMAIC methodology is that it forces improvement project teams to be disciplined about their approach; the insistence on thorough analysis that links cause and effect prevents the all-too-natural tendency to jump straight from problem to solution without sufficient testing of efficacy. This point having been made, it is perhaps equally important to avoid the phenomenon that is known colloquially as 'analysis paralysis', in which an illusory impression of progress is given without any actual progress being made. Action has to be taken, and the next phase of the DMAIC methodology addresses that need.

ACTIVITY 4.7 ....

Construct an input–output diagram for the analysis phase. ●

Phase 4: Improve
The aims of the improvement phase of DMAIC are to:
- generate a small number of solutions that will produce the required changes to relevant processes
- test the solutions
- choose the solution that best meets the requirement

- pilot the solution
- refine the chosen solution, if necessary
- implement the solution fully.

These aims can be translated into the following six-step process.

1 Generate solutions

Although DMAIC is a systematic approach to process improvement, there is still considerable room for insights and creativity, especially during the improvement phase. Good sources for ideas include:

Experience. With experience people develop a good feel for what will work and what will not. Past instances of similar problems and how they were solved can be recalled.

Creativity and idea generation techniques. Although experience can be useful it can result in becoming trapped by the past. This is especially dangerous when new challenges and customer requirements are posing new problems and demanding new processes and methods of operation. In these circumstances, creativity and idea generation techniques such as those you met in Block 3 can be used to overcome the dangers associated with relying on experience.

Research. One way of building up a store of good ideas and at the same time counteracting the restricting effects of experience is to find out what other organisations are doing, perhaps by undertaking a benchmarking exercise.

Modelling and simulation. Modelling and simulation can stimulate ideas as well as predict the outcomes of a particular proposal.

Experts. It is always worth consulting experts, particularly those involved in the day-to-day management and operation of the processes that are the focus of the project. You will recall that Figure 4.27 suggested that Six Sigma project teams need to listen carefully and closely to the voices of the process and the people. This is an important aid to pinpointing problems in the system; it can also be an important way of generating solutions.

All these sources of ideas have their place in generating solutions but it is often the case that a solution, or the main features of it, will become increasingly clear during the preceding phases. It is only in rare instances that 'off the wall' or 'out of the box' thinking is required.

The proposed solutions have to be developed to a point where they can be tested in order to verify that they will correct the problem or realise the opportunity and that they are acceptable to the stakeholders in the process.

2 Test the solutions

Possible solutions have to be tested:

- technically – will they do the job?
- organisationally – are they acceptable?
- financially – are they cost-effective?

The technical suitability of a solution will not simply depend on its output performance; more mundane issues may also be important. Proposed solutions will often involve some alterations to, or rearrangement of, transformation processes and frequently the changes will have physical implications that will also need to be modelled and tested. Once the solutions have been ranked technically it is important that they are tested for acceptability.

The acceptability of a solution to the people who will have to operate it or live with its consequences on a day-to-day basis is at least as important as its technical suitability. There is no point in pushing ahead with the implementation of a solution that, while technically brilliant, is unacceptable organisationally. If the opinions of stakeholders in the process have been gathered assiduously during previous phases of DMAIC the project team will probably have developed a good understanding of what will be acceptable. This should not, however, be used as a reason for not testing acceptability explicitly.

Some explanation and promotion of the various solutions that are being considered is a normal feature of acceptability testing. Most people have an instinctive resistance to change and need to be persuaded of its value. Care must be taken to ensure that this does not become over-selling.

The final test of the proposed solution involves a financial appraisal of its costs and benefits and an assessment of whether any investment in the solution is worth making. Costs, benefits and their timings should be honestly assessed and various measures applied to them. Many projects have the aim of reducing the waste in a system; the elimination of waste will always be financially beneficial unless investment or other costs exceed benefits, but attempts to justify Six Sigma projects are often frustrated by the organisation's approach to financial appraisal. In many organisations there is an over-reliance on payback as a measure of project investment. Payback is a simple approach that is easy to understand and apply. It involves calculating the period of time that it takes the positive cash flows associated with a project to recoup any initial capital investment. Figure 4.36 illustrates the principle.

Payback has an alluring simplicity but, as Figure 4.36 illustrates, it suffers from serious flaws. Curve B has a shorter payback period (point x) than curve A (point y) and is to be preferred if payback alone is the decision criterion. But the cumulative cash flows modelled by curve A indicate that although it requires more initial investment it may be the better project. Using payback as a measure does not allow the whole life of the project to be evaluated since it ignores the size, pattern and timing of cash flows that occur after the payback point.

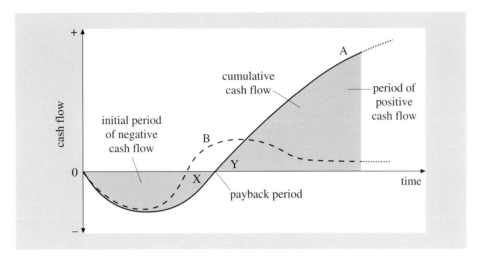

Figure 4.36   Payback as a way of measuring the financial attractiveness of a project

More sophisticated approaches to the financial modelling of project assessment, such as those based on discounting, are often employed to overcome the limitations of payback. Attempts to apply internal rate of return and net present value are often frustrated by organisations' use of a single 'hurdle' or discount rate. Six Sigma projects will involve the organisation in different risks. For example, a project that involves a novel service or product to be sold into a new market and that relies on significant technological innovation will be riskier than replacing an old machine with a new one. And in between these two extremes lie a range of different risk exposures.

It is a fact of human nature that the riskier the undertaking, the more return it must offer to attract investment. The outsiders in a horse race are offered at longer odds (a better price) than those horses that have more chance of winning – the favourites. One measure of the risk that business investors demand is the returns on stock markets. Historically, these have been on average 7.5 to 8 per cent more than the risk-free returns provided by short-dated (typically 90 day) government bonds. However, the *average* extra amount of return means that businesses that are riskier than the average have to deliver a greater return to keep investors happy. Those that are safer than the average can produce a lower return. The resulting curve of return versus risk is shown in Figure 4.37; it is also known as the capital asset pricing model.

Each of the activities in a company can be regarded as an asset that has to produce a level of return that reflects its riskiness. In this way a company is a bundle of these assets and is analogous to the stock market. Applying the logic of the capital asset pricing model to a single company gives the company-specific model shown in Figure 4.38. This shows the effect of applying a single discount rate to all the project opportunities in a company regardless of their riskiness.

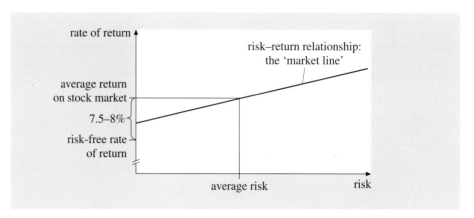

Figure 4.37   The risk–return relationship

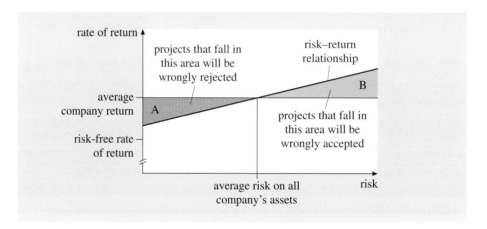

Figure 4.38   The capital asset pricing model applied to a single company

ACTIVITY 4.8 ....

Figure 4.38 suggests that a project that falls into area A in the diagram will be wrongly rejected. Why? Conversely, a project that falls into area B will be wrongly accepted. Why? ●

The implications of the picture shown in Figure 4.38 are that projects with higher than average risk are relatively easy to justify financially – or easier than they should be – but that projects with a lower than average risk are more difficult to justify. The use of a single hurdle rate biases the organisation's decision making in favour of higher risk projects and against operations-focused systems engineering projects, which tend to be of lower risk.

3 Choose the solution that best meets requirements

By this stage of the DMAIC methodology the project team and those that have been involved as stakeholders will have developed a good idea of the solution that they want, and will have tested it for technical and financial feasibility. The final choice of solution can be made by appropriate members of the organisation's senior management team, informed by the project team and the stakeholders. The decision is most often made at a presentation, at which the project process and the outcomes of each of the phases of the methodology can be fully explained.

4 Pilot the solution

There are many reasons for undertaking a pilot of the chosen solution. Some of the most important are to:

- reduce the risk of failure
- avoid misunderstandings or mistakes
- prevent problems and teething troubles
- improve or fine-tune the solution
- learn lessons for the full implementation process
- strengthen the level of acceptance of the solution
- confirm the validity of the costs and benefits estimates
- check that the solution will deliver the anticipated performance benefits
- deliver the solution to a particular work team that is under pressure.

Any change is likely to encounter the 'law of unintended consequences'. This takes various forms but broadly states that any change will have consequences that were not foreseen. These may be welcome, adding to the benefits of the change, or unwelcome; it is the latter that are more often noted. Implementing a pilot of the solution will help to identify and eliminate some of its negative unforeseen consequences, although not, of course and sadly, all of them.

The larger the size of the change that will be implemented, the greater the need for a pilot. Additionally, if the proposed change involves significant investment or is not easy to reverse, a pilot implementation will help to reduce the risks of costly mistakes.

To ensure the success of the pilot it is useful to:

- Define the objectives of the pilot and decide on the measures that will be used to judge its success or failure.
- Decide on the data that will be collected to confirm success or failure and how they will be analysed.
- Undertake a force field analysis to identify the actions that can be taken to promote the solution or remove or reduce opposition to it.
- Identify and plan the activities that will be needed to undertake the pilot. This involves defining the activities that will be required, identifying the resources needed, allocating responsibilities, and estimating durations and timings. The resulting plan will be valuable, additionally, at the full implementation stage.

● Develop a communications plan to ensure that the staff who will be affected by the change are aware that it will be taking place, the reason for making the change, the benefits that are likely to result, how it will affect their work, and so on.

The pilot preparation checklist given in Box 4.7 can be used to ensure that all aspects of the pilot have been considered. The pilot should then be undertaken, the data should be collected and analysed, and any lessons learned should be carried forward to the final steps of implementation.

BOX 4.7   PILOT PREPARATION CHECKLIST

Objectives

1   What are the overall objectives for the pilot?
2   What are the success criteria for the pilot?
3   What are the failure criteria for the pilot?

Authority

4   Who has the authority to allow the pilot to proceed?
5   Who has the authority to stop the pilot?

Documents

6   What documentation will be produced for the pilot?
7   Who will require copies of the documentation?
8   What will happen to documents after the pilot?

Equipment

9   How will the equipment be obtained for the pilot?
10 Will the equipment require modification?
11 Will the equipment need to be specially set up?
12 What will happen to the equipment after the pilot?

Environment

13 What must the environmental conditions be?
14 Will the conditions during [the] pilot need to be recorded?

Operators

15 How many operators will be used for the pilot?
16 How will the operators be selected?
17 Will the operators be typical of those in the future process?
18 What training will be given to the operators?
19 Will a supervisor be needed?

Measuring

20 What will the measurement system be?

21 Will the measuring system require special calibration?

22 What extra measurements will be made in the pilot?

23 What measurements will be repeated during the pilot?

Statistical process control (SPC)

24 Will SPC be used during [the] pilot?

25 What type of SPC will be used and what are the limits?

Samples

26 How many samples will be produced?

27 What will happen to the samples after the pilot?

Records

28 What records need to be kept during the pilot?

(Source: Marconi, 2000, p. 75)

5 Refine the chosen solution if necessary

The pilot will reveal actions that need to be taken to improve the solution, either to make it more effective or to make its implementation easier, and a full review of the lessons that have been learned should be undertaken. This review should consider not only whether the pilot met the success/failure criteria but should also take into account the views of the staff involved. The aim is to develop an agreed set of modifications that incorporate all the lessons that the pilot has taught.

In addition, any issues associated with a need to scale up the pilot should be considered. These can be incorporated into a failure mode and effect analysis. The post-pilot checklist given in Box 4.8 can be used or adapted to ensure that everything has been considered.

BOX 4.8   PILOT REVIEW CHECKLIST

To be used on pilot completion

Objectives

1 Have the overall objectives for the pilot been met?

2 Have the results been fully analysed statistically?

3 Was the pilot a success?

Documents

4   What has been done with the pilot documentation?

5   Will further changes to the documents be needed?

6   Will additional documents need to be produced?

Equipment

7   What has happened to the equipment since the pilot?

8   Did the equipment satisfy the requirements?

9   Are further changes to the equipment required?

Environment

10 Were the environmental conditions adequate?

11 Will additional environmental controls be required?

Operators

12 What were the operators' views of the pilot?

13 Was their training adequate?

14 Can staffing levels be reduced?

15 Will skill levels need to be improved?

Measuring/Statistical Process Control (SPC)

16 Was the measurement system adequate?

17 What changes are needed to the measurement system?

18 Were the Statistical Process Control methodology and the limits adequate?

Samples

19 What has happened to the samples from the pilot?

20 Can they be used, or are they to be scrapped?

Records

21 Will records need to be kept, and for how long?

Future

22 Are further pilot studies required?

23 Who has the authority for agreeing the full roll-out?

24 What needs to be done for the full roll-out of the change?

25 What needs to be done to satisfy ISO 9000, or other similar standards?

(Source: Marconi, 2000, p. 76)

6 Implement the solution fully

The full implementation of the solution can take place once the lessons from the pilot have been incorporated and any additional actions have been completed. The extra activities that need to be undertaken could include:

- full training of staff or the training of additional staff
- turning the draft system documentation into the final version
- developing control plans for the new process
- ordering and installing the extra equipment required.

ACTIVITY 4.9 . . . .

Construct an input–output diagram for the improvement phase. ●

Phase 5: Control

The new system must be controlled in such a way that it delivers the anticipated benefits in a consistent manner, day after day. Other techniques are likely to be needed to ensure that the system is kept up to the mark. Examples include SPC and the poka-yoke method for the prevention of defects.

The new process can be handed over fully to the operations team once the project team is convinced that the new process is under control and is performing consistently. The Six Sigma team will have learned a great deal during the various phases of the project. It is important that this knowledge is shared with other areas so that other teams may learn from it and not repeat any mistakes that may have been made.

ACTIVITY 4.10 . . . .

Construct an input–output diagram for the control phase. ●

7.2 Evaluating Six Sigma

There is no doubt that the Six Sigma approach has many adherents. Many strengths are claimed for it, and the source of most is attributed to Six Sigma's industrial roots and its emphasis on projects being undertaken only if there is a strong expectation that they will be economically viable. Brussee is one of the relatively few authors of books on Six Sigma that are more measured. He says:

> Six Sigma is similar to earlier quality programs such as Total Quality Control (TQC) and ISO 9000. The biggest differences are that Six Sigma is packaged better, is tied closely to bottom-line profits, and has gotten a high degree of support from top levels of management.

> (Brussee, 2006, pp. xi–xii)

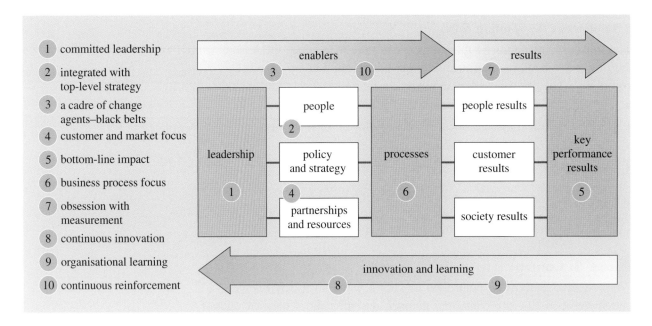

Figure 4.39   The Excellence Model and Six Sigma (Source: Oakland and Marosszeky, 2006, p. 272)

Instead of likening TQM and ISO 9000, Oakland and Marosszeky (2006) draw comparison between Six Sigma and the EFQM Excellence Model. Figure 4.39 shows their mapping of Six Sigma on to the Excellence Model.

Now read Offprint 9 for a fuller evaluation of the pros and cons of Six Sigma.

After you have read the offprint watch Programme 3 *Problem Solving in Action: Six Sigma at ScottishPower.*

8   JUST IN TIME/LEAN PRODUCTION

8.1   Just in time

One of the ways in which manufacturers have sought to improve their performance is by exercising greater control over their suppliers and the internal movement of work in progress. Just in time (JIT) is one of the mechanisms they have used to achieve this.

JIT originated in Japan. It is essentially a production control mechanism, dependent largely on working practices, in which the start of a production run is triggered by end-product demand. Rand (1990) identified its key features as follows:

1   JIT is a pull system, so demand for the final product pulls production through the factory.

2   Production is controlled by cards known as kanbans which indicate when an operation is ready to pull production from the previous operation.

3   JIT encourages the reduction of batch size and set-up time. The former leads in turn to improvements in quality because it enables defects to be detected and corrected quickly.

4   Improved quality is both a requirement for and a benefit of JIT, with quality control being achieved through the use of total inspection, in which every worker is responsible for checking all of his or her production, and through the use of quality failure prevention techniques.

5   It demands employee participation.

The benefits that are claimed for JIT include: reduction in the amount of inventory, better space utilisation, requirement for fewer storage facilities, and improved customer service. (There is a danger, of course, that a powerful customer may simply shift inventory to less powerful suppliers in the environment. If real reductions are to take place, JIT has to be implemented through the entire supply chain.)

JIT would probably have been included in the Techniques block of this course if it were not for two things. The first is its link to lean production – indeed, lean production is sometimes referred to as 'big JIT'. The second is JIT's ability to act as a trigger for change, driving improvement through the system. It can do this because attempts to introduce it are almost bound to reveal any weaknesses such as bottlenecks in a production system, especially if the processes are complicated and there are a number of possible routes through them.

Harrison (1994) suggests that JIT should be implemented in a planned series of stages:

1   Examine the fit between JIT and the business strategy.

2   Make sure basic disciplines such as cleanliness, tidiness and concern for safety and quality are in place.

3   Bring processes under control.

4   Eliminate waste.

5   Eliminate errors at source through the use of error-proof devices and automatic condition-monitoring of machines.

8.2   Lean production

Although its origins can be traced back to Henry Ford's production system, lean production (LP) is very strongly associated with Japan, and in particular with the Toyota Motor Company. The concept that unites JIT and LP is waste, or *muda* as it is known in Japan. Lean thinking has been developed as 'a powerful antidote to *muda*' (Womack and Jones, 2003, p. 15).

ACTIVITY 4.11 . . . .

Without looking back at Box 4.6, identify four different types of waste that might be generated in a manufacturing or a service organisation. ●

The five lean principles are: value; value stream; flow; pull; perfection.

I shall look at each of them in turn, citing from Womack and Jones (2003).

Value –

> The critical starting point for lean thinking is *value*. Value can only be defined by the ultimate customer. And it's only meaningful when expressed in terms of a specific product (a good or a service, and often both at once) which meets the customer's needs at a specific price at a specific time.
>
> Value is created by the producer.
>
> (p. 16)

> [S]pecifying value accurately is the critical first step in lean thinking. Providing the wrong good or service the right way is *muda*.
>
> (p. 19)

Value stream – If you are familiar with Porter's concept of the value chain (Porter, 1985) you should note that Womack and Jones distinguish between it and their concept of the value stream. They see Porter's concept as aggregating activities like production, marketing and sales across a range of products.

> The *value stream* is the set of all the specific actions required to bring a specific product ... through the three critical management tasks of any business: the *problem-solving task* running from concept through detailed design and engineering to product launch, the *information management task* running from order-taking through detailed scheduling to delivery, and the *physical transformation task* proceeding from raw materials to a finished product in the hands of the customer. Identifying the *entire* value stream for each product ... exposes enormous, indeed staggering, amounts of *muda*.
>
> Specifically, value stream analysis will almost always show that three types of actions are occurring along the value stream: (1) Many steps will be found to unambiguously create value: welding the tubes of a bicycle frame together or flying a passenger from Dayton to Des Moines. (2) Many other steps will be found to create no value but to be unavoidable with current technologies and production assets: inspecting welds to ensure quality and the extra step of flying large planes through the Detroit hub en route from Dayton to Des Moines (we'll term these Type One *muda*). And (3) many additional steps will be found to create no value and to be immediately avoidable (Type Two *muda*).
>
> (pp. 19–20)

Flow – After the value stream has been mapped fully and steps that are clearly wasteful have been eliminated it is necessary to make all the remaining value-creating steps flow.

Pull – Customers should pull goods and services from producers rather than producers pushing them.

Perfection – An organisation moves towards perfection as the other four principles interact with each other in a virtuous circle:

> Perhaps the most important spur to perfection is *transparency*, the fact that in a lean system everyone ... can see everything, and so it's easy to discover better ways to create value.
>
> (p. 26)

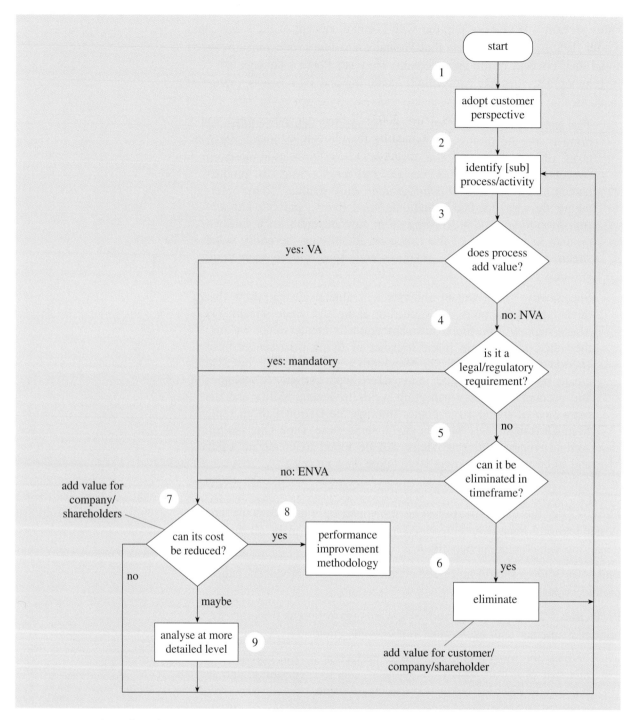

Figure 4.40   Flow chart for separating processes and activities that can be eliminated from those that cannot

In theory you might assume that perfection implies the elimination of all waste but some activities absorb resources without creating value that cannot be eliminated. These are termed essential but non-value-added (ENVA). Often these are processes or activities associated with legal or regulatory requirements such as the US Sarbanes-Oxley Act (also known as the Public Company Accounting Reform and Investor Protection Act). Sometimes it is difficult to distinguish whether activities add value or not. For example a glossary of lean terms drawn up by an NHS hospital trust suggests that 'in air travel, [an] essential but non value added activity would be security checks' (Bolton Hospitals NHS Trust, 2002). This is debatable when you consider that 'value can only be defined by the ultimate customer' (Womack and Jones, 2003, p. 16). A flow chart that can help to separate those processes and activities that can be eliminated from those that cannot is shown in Figure 4.40.

Implementing lean production

A typical organisation chart for a lean business is shown in Figure 4.41. Following Womack and Jones, the boxes in the figure have been drawn so that their sizes are proportional to the number of employees represented by each. The 'lean promotion' function includes the process mappers, logistics support for improvement teams and trainers in lean methods, all of whom should report to a change agent.

Table 4.3 shows the timetable for implementation that Womack and Jones suggest. As you can see, they are recommending a five-year timescale.

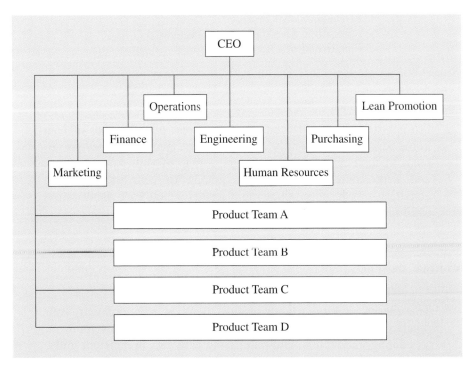

Figure 4.41   Prototype lean organisation (Source: Womack and Jones, 2003, p. 257)

Table 4.3   Time frame for the lean leap

Phase	Specific steps	Time frame
Get started	Find a change agent Get lean knowledge Find a lever Map value streams Begin *kaikaku* [a rapid change event that delivers radical improvements] Expand your scope	First six months
Create a new organization	Reorganize by product family Create a lean function Devise a policy for excess people Devise a growth strategy Remove anchor-draggers Install a 'perfection' mind-set	Six months through year two
Install business systems	Introduce lean accounting Relate pay to firm performance Implement transparency Initiate policy deployment Introduce lean learning Find right-sized tools	Years three and four
Complete the transformation	Apply these steps to your suppliers/customers Develop global strategy Transition from top-down to bottom-up improvement	By end of year five

(Source: Womack and Jones, 2003, p. 270)

As Womack and Jones (2003) point out:

> [To get started,] you'll need a change agent plus the core of lean knowledge (not necessarily from the same person), some type of crisis to serve as a lever for change, a map of your value streams, and a determination to *kaikaku* quickly to your value-creating activities in order to produce rapid results which your organization can't ignore.
>
> (p. 247)

Womack and Jones (2003) also give the following advice on implementation:

1   Identify product families

This means identifying your product families and rethinking your functions to realign marketing/sales, product development, scheduling, production, and purchasing activities in coherent units.

(p. 256)

2 Introduce lean promotion function

An even better idea is to combine your quality assurance function with your lean promotion function so that quality enhancement, productivity improvement, lead-time reduction, space savings, and every other performance dimension of your business are considered equally and simultaneously.

(pp. 256, 257)

3 Reduce input of labour early on

Our rule of thumb is that when you convert a pure batch-and-queue activity to lean techniques you can eventually reduce human effort by three quarters with little or no capital investment. When you convert a 'flow' production setup ... to lean techniques, you can cut human effort in half. ... [I]n product development and order-taking, converting from batch-and-queue to flow will permit your organization to do twice the work in half the time with the same number of people.

(pp. 257, 258)

Womack and Jones (2003) give two alternatives for reducing the labour input, depending on whether the crisis is severe or not. If the crisis is severe:

The correct thing to do is to face it up front, by estimating the number of people needed to do the job the right way, and moving immediately to this level. Then you must guarantee that no one will lose their job in the future due to the introduction of lean techniques. And you must keep your promise.

(p. 258)

If there is no significant crisis then it is essential to find alternative work for those who are released from their current positions. The best way of doing this is:

to devise a growth strategy which absorbs resources at the rate they are being freed up.

(p. 259)

8.3 Evaluating lean production

Now read Offprint 10. It provides further information on lean organisations and evaluates the lean production approach, comparing and contrasting it with Six Sigma.

9 CONCLUSION

Earlier in the course I looked at three generic methods for problem solving and improvement, and in this block I have described a variety of more specialised and more sophisticated approaches. They fall into three types:

1   quality organisation led – ISO and EFQM
2   systems approaches
3   industry led – Motorola and Toyota.

As you have seen, all these approaches have strengths and weaknesses. They also have forerunners, variants and imitators, and at some point in time radically different approaches are likely to emerge to challenge their claims to be included in this course. However, for the time being at least, the approaches that have been covered can all claim to offer something that separates them from passing fads.

The same cannot be said of all approaches, however, as you can see in Boxes 4.9 and 4.10. Box 4.9 is the transcript of part of a radio interview with David Craig, whom the *Observer* describes as a 'consultant turned whistleblower' (Cohen, 2006). Box 4.10 looks at business process re-engineering. This is a radical approach to improvement that appeared in 1990 (Davenport and Short, 1990; Hammer, 1990) and was swiftly taken up very enthusiastically. However, by 1995 Davenport himself was saying 'To most business people in the United States, re-engineering has become a word that stands for restructuring, layoffs and too-often failed change programs' (Davenport, 1995, p. 1).

BOX 4.9 TRANSFORMING THE ORGANISATION

I was part of the group about ten years ago working for a company called Gemini Consultancy which was part of Capgemini.

We wanted to sell bigger projects, lots of consultancies and big IT systems. We came up with this idea 'organisational transformation'. We published a book called *Transforming the Organisation* [Gouillart and Kelly, 1994] ... When our book came out the *Economist* reviewer [*Economist*, 1995] said that any client stupid enough to buy this concept of transformation would find that he was employing an army of consultants for a century. We made quite a lot of progress in using this word transformation to sell big, big consulting projects to private companies but the thing really took off under New Labour who adopted the transformation concept hook line and sinker ... always talking about transforming education, transforming the health service.

My message for them would be 'guys, we didn't mean it seriously'. Nobody actually needs to buy transformation. It was only a trick, only a way of selling lots and lots of consulting with lots of IT systems behind it.

(David Craig, speaking on *Word of Mouth*, broadcast on BBC Radio 4, 11 September 2006)

BOX 4.10 REENGINEERING REVISITED: WHAT WENT WRONG WITH THE BUSINESS-PROCESS REENGINEERING FAD. AND WILL IT COME BACK?

At its height, business-process reengineering was one of the biggest business ideas ever. Business historians of the future will characterize the 1990s as the decade of reengineering. Described in more than 25 books, featured in articles in every major business publication, discussed at hundreds of conferences, reengineering penetrated every continent, with the possible exception of Antarctica.

Reengineering became a money machine for several of its constituents: the gurus who propounded the idea (at least a couple of them!), the consulting firms that offered reengineering services to their clients, and the software vendors who managed to convince firms that their wares were critical to successful reengineering. Unfortunately, the idea didn't enrich those who were most responsible for its birth and continued life within organizations: the faithful practitioners. These individuals played their customary heroic roles on the reengineering stage, but others got all the credit.

Reengineering Defined

Reengineering today means different things to different people. In the early writings on reengineering, however, there was substantial consensus that reengineering incorporated the following ideas:

- Radical redesign and improvement of work.
- Attacking broad, cross-functional business processes.
- 'Stretch' goals of order-of-magnitude improvement.
- The use of IT as an enabler of new ways of working.

In the general business audience, however, other meanings proliferated. To some, reengineering came to mean any attempt to change how work is done – even incremental change of very small processes. To others, it became a code word for downsizing. The latter meaning wasn't really fair, since none of the original literature on reengineering had stressed that objective. It was a somewhat cynical adoption of the word by senior executives (and their communications staffs) that brought this meaning into being.

We still remember the first time the reengineering-as-downsizing notion appeared in the press. In 1995, Pacific Bell announced that it was cutting 10,000 employees. Because of 'reengineering', it didn't need them anymore, the press release said. We were conducting some research at Pac Bell at the time, and we knew that although the company was doing some reengineering, it certainly wasn't far enough along for anyone to know how many people (if any) would be freed up. Shortly thereafter, Apple Computer Inc. announced a similar reduction using the 'R word'. We were also familiar enough with that company to know that it wasn't true reengineering.

What happened to the term reengineering is typical of the proliferation of meanings that accompanies any successful new business idea. Consultants, middle and senior managers, and vendors had lots of incentives to jump on the reengineering bandwagon. Experts in continuous improvement, systems analysis, industrial engineering and cycle-time reduction all suddenly became experts in reengineering. We once heard a staffer from the California legislature say that reengineering was 'any project I want to get funded'.

Of course, saying that all these diverse activities were forms of reengineering raised expectations for the concept and no doubt hastened its demise. The late adopters of the term dropped it rapidly as soon as it became unpopular.

Where did reengineering go astray? Like any other business idea, reengineering had to be bought by companies and sold by business gurus. The failure of reengineering can be attributed to both parties.

Guru Shortcomings

A key factor in the rise of reengineering was Michael Hammer's 1990 *Harvard Business Review* article on reengineering ('Don't Automate, Obliterate') and the subsequent book Hammer co-authored with Jim Champy, *Reengineering the Corporation*. These two sources made reengineering look both appealing and easy. But in late 1996, a front-page *Wall Street Journal* article featured a confession by Hammer: 'Dr. Hammer points out a flaw: He and the other leaders of the $4.7 billion re-engineering industry forgot about people. "I wasn't smart enough about that," he says. "I was reflecting my engineering background and was insufficiently appreciative of the human dimension. I've learned that's critical."'

Hammer's earlier rhetoric certainly neglected the human element, with phrases such as, 'In reengineering, we carry the wounded and shoot the stragglers,' and, 'It's basically taking an ax and a machine gun to your existing organization.' This rhetoric not only made employees fear for their livelihoods; it also raised expectations of managers for revolutionary changes that couldn't be delivered.

There is little doubt that the Hammer and Champy version of reengineering was guilty of overblown rhetoric. (Most of this rhetoric comes either from Hammer or from ghostwriters; Champy is more genial and mild-mannered.) Both in Hammer's 1990 *Harvard Business Review* article and in Hammer and Champy's 1993 book, the claims were extravagant and unsupported by fact. The book cover suggests, for example, that 'everything you thought you knew about business is wrong', and highly simplistic arguments are made throughout.

The greatest shortcoming of the Hammer and Champy reengineering work is not that it neglects people or that it employs overblown rhetoric, but that it fails to acknowledge how difficult, time-consuming and expensive it is to reengineer. They implied that one could reengineer an entire corporation in as little as a year. We aren't suggesting that Hammer and Champy intentionally misled anyone; they're both honorable men. This unfounded optimism was, however, a major factor in the rapid rise and fall of reengineering. Inspired by the book, managers initiated projects with high expectations of rapid success. But when they encountered difficulties and slipping project deadlines, many became disenchanted and dropped their projects.

Implementation Problems

The gurus of reengineering made some mistakes, but then so did many of the practitioners. First of all, many managers reinforced the numerous errors made by the gurus. They focused too much on process and not enough on practice – and didn't involve the people who did the work. Just as some irresponsible writers and consultants generated exaggerated rhetoric and repackaged ideas in reengineering, so did companies.

Corporate communications departments teamed with senior executives to create reengineering programs with names like Advantage 2000, Program 10X and, perhaps most unfortunately, Project Infinity. They predicted radical improvements well before they were achieved. In some cases, they argued for funding by calling their projects 'reengineering'. The worst offense, of course, was to lay off people and dignify the act as reengineering.

But there were several other, more subtle problems with how companies implemented reengineering. One is that executives turned too much of their initiatives over to outside firms – both consultants and vendors of enterprise software (such as ERP systems). The software vendors supplied a relatively easy way to automate a broad range of business processes in an integrated fashion.

Not surprisingly, the managers of reengineering projects flocked to the enterprise software vendors such as SAP AG, Oracle Corp. and PeopleSoft Inc. and wrapped up their reengineering and ERP projects into one integrated change program. But these companies probably relied too heavily on the software as the way to implement reengineering. While the packages were built around best practices, they were generic rather than specific to a particular company's needs. Because it was difficult to modify these systems, most firms ended up with the same processes and information support as every other firm in their industry. Reengineered processes were supposed to yield competitive advantage, but this was impossible with heavy reliance on an enterprise software package.

Also, many corporate reengineers took on too much change at once. Encouraged by the rhetoric of some gurus, they tried to change multiple processes, information systems, organizational structure and sometimes even business strategy all at once. Such all-encompassing change in a short time frame is difficult, if not impossible. One observer noted that it was akin to pilots attempting to change all the engines in a jet airplane at once while flying through the sky. It may be possible, but the risk of failure is great.

The Good Stuff

Despite the problems with the reengineering movement, you might ask, does it have any ideas worth keeping? You bet.

In fact, almost all the ideas within reengineering have substantial merit when used in moderation. Certainly, firms should sometimes address broad, cross-functional processes. And from time to time, they need a serious kick in the pants.

Sometimes it's better to throw a broken process away altogether and start from scratch than to improve it incrementally. And IT can no doubt be a powerful enabler and reshaper of processes.

The key is to also remember that reengineering involves risk. Any time an organization needs radical change to deliver the results it needs, it's more likely to fail or to come up short.

Like baseball players who swing for the fences or soccer stars who take kicks from midfield, they'll miss their goal most of the time. But sometimes, desperate moves are called for.

(Source: Davenport et al., 2003a, excerpted from Davenport et al., 2003b)

Selecting which approach to use is seldom straightforward. You will need to make that choice for your project and, more importantly, you are likely to want to make that choice 'for real' at some point, so selection is the first topic I shall look at in Block 5.

REFERENCES

Ackoff, R. L. (1979) 'The future of operational research is past', *Journal of Operational Research Society*, Vol. 30, pp. 93–104.

Ackoff, R. L. (1981) 'The art and science of mess management', *Interfaces*, Vol. 11, pp. 20–6.

Bolton Hospitals NHS Trust (2002), *Bolton Improving Care System (BICS) – Lean Glossary of Terms*, http://www.boltonhospitals.nhs.uk/publications/ bics/leanglossary.html (accessed 4 July 2007).

BS 5750:1979 *Quality Systems*, London, British Standards Institution.

BS EN ISO 9000:2000 *Quality Management Systems. Fundamentals and Vocabulary*, London, British Standards Institution.

BS EN ISO 9000:2005 *Quality Management Systems. Fundamentals and Vocabulary*, London, British Standards Institution.

BS EN ISO 9001:2000 *Quality Management Systems. Requirements*, London, British Standards Institution.

BS EN ISO 9001:2008 *Quality Management Systems. Requirements*, London, British Standards Institution.

BS EN ISO 9004:2000 *Quality Management Systems. Guidelines for Performance Improvements*, London, British Standards Institution.

BS EN ISO 9004:2009 *Managing for the sustained success of an organization. A quality management approach*, London, British Standards Institution.

Brussee, W. (2006) *All About Six Sigma*, McGraw-Hill, New York.

Bulow, I. von (1989) 'The bounding of a problem situation and the concept of a system's boundary in Soft Systems Methodology', *Journal of Applied Systems Analysis*, no. 16, pp. 35–41.

Checkland P. B. (1972) 'Towards a systems-based methodology for real-world problem solving', *Journal of Systems Engineering*, Vol. 3, pp. 87–116.

Checkland, P. B. (1981) *Systems Thinking, Systems Practice*, 1st edn, Chichester, Wiley.

Checkland, P. B. (1999) *Systems Thinking, Systems Practice*, 2nd edn, Chichester, Wiley.

Checkland, P. B. and Scholes, J. (1990) *Soft Systems Methodology in Action*, Chichester, Wiley.

Churchman, C. W. (1971) *The Design of Inquiring Systems*, New York, Basic Books.

Cohen, N. (2006) 'Labour's NHS is a real tonic for the Tories', *Observer*, 31 December, p. 12.

Davenport, T. H. (1995) 'The fad that forgot people', *Fast Company*, no. 1, November.

Davenport, T. H. and Prusak, L. with Wilson, H. J. (2003a) 'Reengineering revisited – what went wrong with the business-process reengineering fad. And will it come back?', *Computerworld*, 23 June, http://www.computerworld.com/managementtopics/management/story/0,10801,82290,00.html (accessed 4 July 2007).

Davenport, T. H. and Prusak, L. with Wilson, H. J. (2003b) *What's the Big Idea: Creating and Capitalizing on the Best Management Thinking*, Harvard, Harvard Business School Press.

Davenport, T. H. and Short, J. E. (1990) 'The new industrial engineering: information technology and business process redesign', *Sloan Management Review*, Vol. 31, no. 4, pp. 11–27.

Deming, W. E. (1953) 'On the distinction between enumerative and analytical studies', *Journal of the American Statistical Association*, Vol. 48, no. 262, pp. 244–55.

Deming, W. E. (1975) 'On probability as the basis for action', *American Statistician*, Vol. 29, no. 4, pp. 146–52.

DES (1983) *Science Report for Teachers – 1: Science at Age 11*, London, HMSO.

DES (1985) *Science 5–16: a Statement of Policy*, London, HMSO.

DES (1990) *Standards in Education 1988–89*, London, HMSO.

DTI (1992) *A Positive Contribution to Better Business. An Executive's Guide to the Use of the UK National Standard and International Standard for Quality Systems*, London, Department of Trade and Industry.

The Economist (1995) 'A brief theory of everything', Vol. 335, no. 7913, p. 63.

EFQM (2009a) *EFQM Excellence Model 2010*, Brussels, European Foundation for Quality Management.

EFQM (2009b) *EFQM Transition Guide*, Brussels, European Foundation for Quality Management; http://www.efqm.org/en/PdfResources/Transition_Guide.pdf (accessed 2 February 2011).

Fortune, J. and Peters, G. (1995) *Learning from Failure – the Systems Approach*, Chichester, Wiley.

Fortune, J. and Peters, G. (2005) *Information Systems: Achieving Success by Avoiding Failure*, Chichester, Wiley.

Fortune, J., Peters, G. and Rawlinson-Winder, L. (1993) 'Science education in English and Welsh primary schools: a systems study', *Journal of Curriculum Studies*, Vol. 25, pp. 359–69.

Gouillart, F. and Kelly, J. (1994) *Transforming the Organization*, New York, McGraw-Hill.

Hammer, M. (1990) 'Re-engineering work: don't automate, obliterate', *Harvard Business Review*, July–Aug., pp. 104–12.

Harrison, A. (1994) 'Just-in-time manufacturing', in Storey, S. (ed.) *New Wave Manufacturing Strategies: Organizational and Human Resource Management Dimensions*, London, Paul Chapman, pp. 177–203.

Imai, M. (1997) *Gemba Kaizen. A Commonsense, Low-Cost Approach to Management*, New York, McGraw-Hill.

ISO (1998) *Introduction to the Revision of the ISO 9000 Standards*, International Organization for Standardization, http://www.iso.ch/presse/intro9k.htm (accessed 22 June 2000).

ISO (2010a) *About ISO*, International Organization for Standardization, http://www.iso.org/iso/about.htm (accessed 2 February 2010).

ISO (2010b) *What's different about ISO 9001 and ISO 14001*, International Organization for Standardization, http://www.iso.org/iso/about/discover-iso_whats-different-about-iso-9001-and-iso-14001.htm (accessed 2 February 2010).

ISO, *ISO 9000/ISO 14000 – ISO 9000:2000 Selection and Use*, http://www.iso.org/iso/en/iso9000-14000/understand/selection_use/maintaining.html (accessed 27 July 2007).

Jenkins, G. M. (1969) 'The systems approach', *Journal of Systems Engineering*, Vol. 1, pp. 3–49.

Marconi (2000) *Marconi 6S Process*, Coventry, Marconi Communications.

Mansell, G. (1993) 'The failures method and soft systems methodology', *Systemist*, Vol. 15, pp. 190–204.

Mansell, G. (1996) Book review, *Systems Research*, Vol. 13, no. 4, pp. 501–3.

Meadows, D. H., Meadows, D. L., Randers, J. and Behrens III, W.W. (1974) *The Limits to Growth*, Universe Books, New York.

Mitev, N. (2000) 'Toward social constructivist understandings of IS success and failure: introducing a new computerized reservation system', *Proceedings of 21st International Conference on Information Systems, Brisbane, Queensland, Australia*.

Oakland, J. (2005) 'From quality to excellence in the 21st century', *Total Quality Management*, Vol. 16, no. 8–9, pp. 1053–60.

Oakland, J. and Marosszeky, M. (2006) *Total Quality in the Construction Supply Chain*, Oxford, Butterworth-Heinemann.

Ohno, T. (1988, originally published in Japan 1978) *Toyota Production System*, Portland, Productivity Press.

Pande, P. S., Neuman, R. P. and Cavanagh, R. R. (2000) *The Six Sigma Way: How GE, Motorola, and Other Top Companies Are Honing Their Performance*, McGraw-Hill, New York.

Peters, G. and Fortune, J. (1992) 'Systemic methods for the analysis of failure', *Systems Practice*, Vol. 5, pp. 529–42.

Porter, M. (1985) *Competitive Advantage*, New York, Free Press.

Quality Progress (1992) Glossary, February.

Rand, G. K. (1990) 'MRP, JIT and OPT', in Hendry, L. C. and Eglese, R. W. (eds) *Operation Research Tutorial Papers*, Operation Research Society, pp. 103–36.

Ryan, T. P. (1989) *Statistical Methods for Quality Improvement*, New York, Wiley.

Snee, R. D. and Hoerl, R. W. (2005) *Six Sigma Beyond the Factory Floor*, New Jersey, Pearson Prentice Hall.

The Open University (2004) T306 *Managing complexity: a systems approach*, Block 2, 'Managing and learning with Information Systems', Part 3 'Using the soft systems method for managing Information Systems on the Taurus Project', Milton Keynes, The Open University.

Womack, J. P. and Jones, D. T. (2003) *Lean Thinking*, London, Simon & Schuster.

ANSWERS TO ACTIVITIES

Activity 4.1

BS EN ISO 9004:2009, p. 30 identifies the maturity levels as follows:

Level 2: Processes are in place to communicate with, select, evaluate, re-evaluate and rank suppliers.

Level 3: Suppliers and partners are identified in accordance with strategic needs or risks. Processes for developing and managing the relationships with suppliers and partners exist.

Level 4: Open communication of needs and strategies occurs with partners.

Activity 4.2

Broader	Narrower
EFQM – building partnerships	ISO – mutually beneficial supplier relationships
EFQM – succeeding through people	ISO – involvement of people
EFQM – adding value for customers and achieving balanced results	ISO – customer focus

Activity 4.3

See Figures 4.42(a) and (b).

Activity 4.4

Two primary tasks are the collection of packets and parcels and the delivery of packets and parcels.

Figure 4.42(a)   A rich picture of the police situation before the introduction of SPC

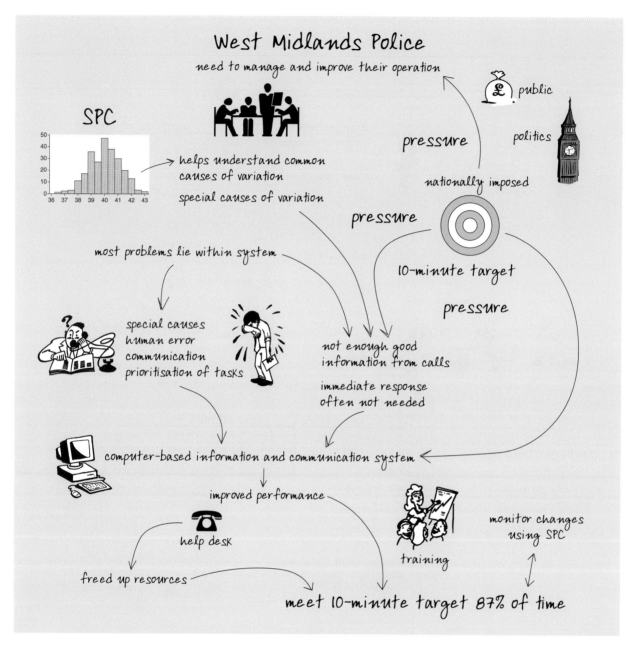

Figure 4.42(b)   A rich picture of the police situation since the introduction of SPC

Activity 4.5

My analysis of the definition phase is shown in Figure 4.43.

Activity 4.6

My analysis of the measuring phase is shown in Figure 4.44.

Activity 4.7

My analysis of the analysis phase is shown in Figure 4.45.

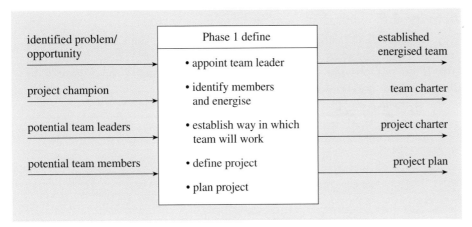

Figure 4.43   DMAIC Phase 1 input–output analysis

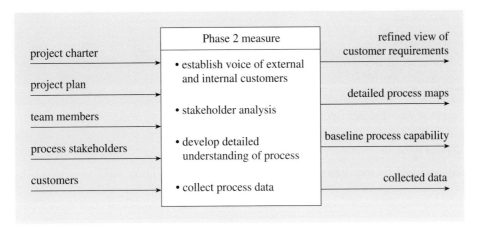

Figure 4.44   DMAIC Phase 2 input–output analysis

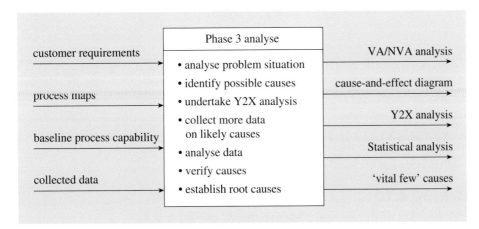

Figure 4.45   DMAIC Phase 3 input–output analysis

Activity 4.8

Projects that fall into area A provide a greater return than the capital asset pricing model suggests is needed for the level of risk. Those falling into area B do not provide enough return. Projects X and Y on Figure 4.46 illustrate these conditions.

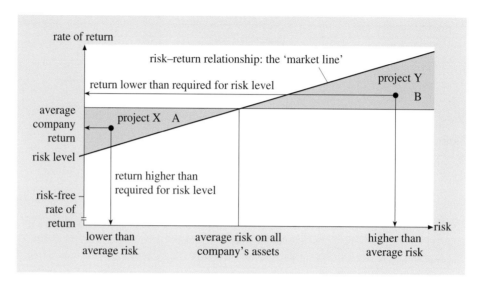

Figure 4.46   Project return and risk

Activity 4.9

My analysis of the improvement phase is shown in Figure 4.47.

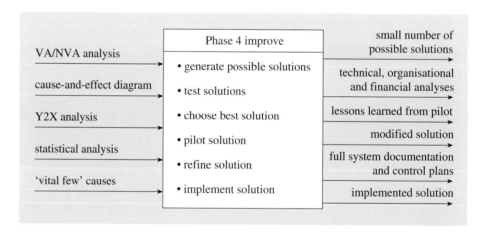

Figure 4.47   DMAIC Phase 4 input–output analysis

Activity 4.10

My analysis of the control phase is shown in Figure 4.48.

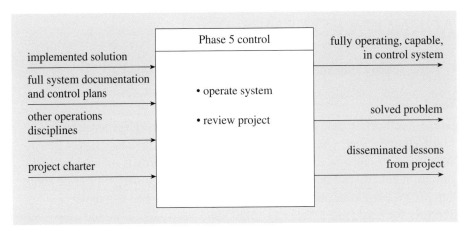

Figure 4.48   DMAIC Phase 5 input–output analysis

Activity 4.11

Toyota executive Taiichi Ohno, whom you met in Section 7.1 and Block 3, has identified the following types of waste (muda in Japanese):

- defects
- overproduction of goods not needed
- inventories of goods awaiting further processing or consumption
- unnecessary processing
- unnecessary movement of people
- unnecessary transport of goods
- waiting.

(Ohno, 1988, pp. 19–20)

ACKNOWLEDGEMENTS

Grateful acknowledgement is made to the following sources:

Text

Box 4.1 & Section 3.1: Oakland, J. (2005) 'Quality to Excellence in the 21st Century'. *Total Quality Management*, Taylor & Francis.; Box 4.2: Permission to reproduce extracts from British Standards is granted by BSI. British Standards can be obtained in PDF format from the BSI online shop: http://www.bsi-global.com/en/Shop/ or by contacting BSI Customer Services for hardcopies: Tel: +44 (0)20 8996 9001, email: mailto:cservices@ bsi-global.com; Activity 4.2: The EFQM Excellence Award – Information Brochure for 2006, © EFQM.; Box 4.10: Davenport, T.H. et al (2003) 'Reengineering Revisited', *What's the Big Idea: Creating and Capitalizing on the Best Management Thinking*. Copyright 2003 by Accenture Ltd and Laurence Prusak, Harvard Business School Press.; Activity 4.11: Womack J.P., and Jones D.T., (1996) *Lean Thinking*, Blackwell Publishing Limited.

Tables

Table 4.2: Pande P.S., Neuman R.P., Cavanagh R.R. (2000) 'An Executive Overview of Six Sigma', *The Six SigmaWay*, © McGraw-Hill Book Co.; Table 4.3 Womack J.P., and Jones D.T., (1996) *Lean Thinking*, Blackwell Publishing Limited.

Figures

Figure 4.1 Oakland, J. (2005) 'Quality to Excellence in the 21st Century'. *Total Quality Management*, Taylor & Francis.; Figure 4.2 and Box 4.2: Permission to reproduce extracts from British Standards is granted by BSI; Figure 4.3: *EFQM Excellence Model*, EFQM and the British Quality Foundation, © 2008 EFQM; Figure 4.4: From *EFQM Transition Guide: How to Upgrade to the EFQM Excellence Model 2010*, © EFQM; Figure 4.39: Oakland J., and Marosszeky M., (2006), *Total Quality in the Construction Supply*, Elsevier Science.; Figures 4.10 and 4.11: Checkland, P. & Scholes, J. (1996) 'The Developed Form of Soft Systems Methodology', *Soft Systems Methodology in Action*, John Wiley & Sons.; Figure 4.12: Checkland, P. (1999) *Soft Systems Methodology: A 30-year retrospective*, John Wiley & Sons.; Figures 4.16, 4.20 & 4.21: Fortune, J. et al (1993) 'Science Education in English and Welsh Primary Schools: A Systems Study', *J. Curriculum Studies*, vol. 25 pp. 363–364, Taylor & Francis Ltd.; Figures 4.17, 4.18 & 4.19: Fortune, J. and Peters, G. (2005) 'The Systems Failures Approach Part 2', *Information Systems: Achieving Success by Avoiding Failure*, John Wiley & Sons.; Figure 4.41 Womack J.P., and Jones D.T., (1996) *Lean Thinking*, Blackwell Publishing Limited.

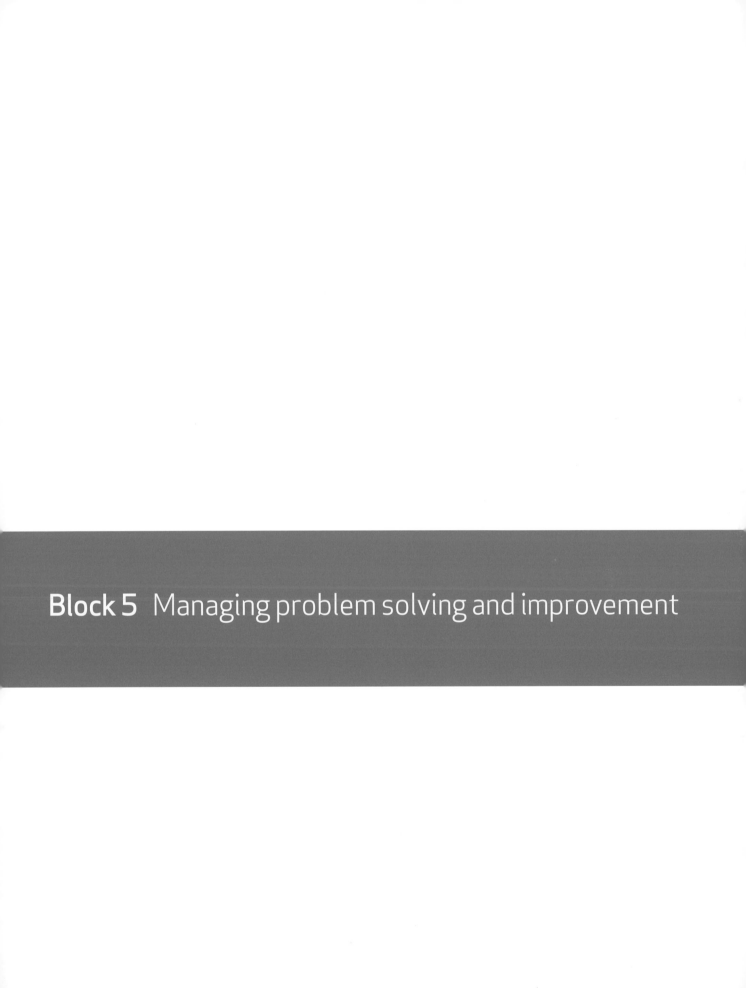

Block 5 Managing problem solving and improvement

CONTENTS

AIMS

The aims of Block 5 are to:

- help you to select an appropriate approach to problem solving and/or improvement for a given context
- set out the advantages and drawbacks of working in problem-solving and improvement teams
- discuss a range of topics associated with organisational climate, culture, leadership and commitment
- introduce a range of measures that can be used to monitor the results of problem-solving and improvement activities
- introduce the topic of portfolio selection and outline techniques that can be used to select projects
- make you aware of reasons why initiatives might fail.

LEARNING OUTCOMES

After studying Block 5 you should be able to:

- appreciate the responsibilities and challenges you face in managing problem-solving and improvement activities
- select an appropriate approach to problem solving and/or improvement for a given context
- appreciate the advantages of working in teams and the problems associated with this method of working
- draw on concepts that are associated with organisational climate, culture, leadership and commitment, and use them to inform your own practice
- devise appropriate measures to monitor the results of problem-solving and improvement activities
- participate in the selection of approaches.

1   INTRODUCTION

There are two aspects to the management of problem solving and improvement. The first is your management of your own problem-solving and improvement projects and activities and the second is your management of the projects and activities undertaken by others. I shall be covering both aspects in this block. Clearly, there is a lot of overlap between them but there are topics such as the management of your own time and the need to gain the commitment of others that relate more to one aspect than the other.

The first topic I shall consider is the selection of approaches to problem solving and improvement. There are three reasons for choosing this as the first topic: it forms a bridge between Blocks 1, 4 and 5; it will be helpful to you in planning your project; and it introduces a number of concepts that are important to Block 5 as a whole.

2   SELECTION OF APPROACHES

It is often the case that an organisation decides to adopt a particular approach to problem solving or improvement for reasons that are almost entirely external to the situation.

ACTIVITY 5.1 . . . .

Suggest three external reasons why an organisation might decide to adopt a particular approach to problem solving or improvement.

However, I am going to assume here that you have a free choice. Which approach should you choose? The answer to this question lies at the intersection of three sets of factors:

1   the characteristics of each of the approaches being considered
2   the characteristics of the problem/opportunity and its context
3   the skills, attitudes, preferences and so on of the person/people who are going to do the work.

The first of these has been taken care of because you are now aware of the characteristics of all the approaches that are covered in this course. Block 1 looked at the characteristics of the problem/opportunity and at some aspects of the context, and I shall now consider context a little further before moving on to the people aspects.

2.1   The context

The two concepts I shall be looking at in this section are organisational climate and culture. Both are somewhat elusive but they are useful notions when trying to understand the processes by which organisational structure and managerial strategies affect people and combine to form the context in which problems and improvement opportunities are located.

The climate of an organisation arises from the interaction between its structure, its size, the technology it uses and so on, and the people who work in it. It determines what the organisation 'feels like' to those who work there. The state of the organisational climate is almost certain to affect both how easily improvement programmes can be introduced and how successful they prove to be in the longer term. As long ago as 1979, Bowey and Carlisle (1979) were arguing that, in order to introduce an innovation such as group working successfully, it is first of all essential to sort out any industrial relations problems. Similarly, a study of the failure of quality circles by Dale and Hayward (1984) noted that the circles were almost bound to collapse if they were introduced when industrial relations in a company were poor.

Climate is a dynamic phenomenon. It changes over time and is affected by the past history of an organisation in that the work experiences of employees in the past are likely to affect the climate in the future. Specifically, perception of the dependability of the organisation – that is, the extent to which it considers the interests of its employees – will critically affect important aspects of the climate such as trust, which in turn will filter down into an atmosphere of cooperation (or non-cooperation, as the case may be).

ACTIVITY 5.2. . . . .

Which part of Block 4 does the last paragraph call to mind?

Clearly, 'getting the climate right' in an organisation is an important precursor to successful problem solving and improvement. But how might this be achieved? This is a very difficult question to tackle. As a starting point it is useful to consider what it is that actually holds organisations together. On what bases does the cooperation between the individuals and interest groups in the organisation rest? Katz and Kahn (1978) argued that companies have three means at their disposal to gain the cooperation of their employees:

1 enforced compliance
2 control of material resources and rewards, such as wages and salaries – usually labelled 'calculative' means
3 manipulation of non-material or symbolic rewards, such as esteem and pleasure in giving service and prestige – usually labelled 'normative' means.

Fortunately, the first of these is not an option that is open to many work organisations. Historically, commercial businesses have tended to rely on calculative means to gain the necessary cooperation of their workforces, while normative means have dominated in voluntary organisations and have also been drawn on to some extent by the public sector. When only calculative means are used, the parties involved (employers and employees) are not usually seeking any involvement over and above that required to ensure that the agreed work tasks are performed: labour (manual and non-manual) is being exchanged for material rewards.

Normative cooperation operates when employees, having taken on at least some of the goals and values that are inherent in the culture of the organisation, cooperate with others to achieve them and take pride in their activities and outputs. It can be encouraged by:

• the use of intrinsically rewarding tasks to develop meaningful and satisfying roles
• the tying together of organisational objectives through small face-to-face groups to generate social pressure towards certain actions
• participation in goal setting so that people see the resulting policies as their own.

The concept of rewards is important in both calculatively and normatively based cooperation. A reward is something that is desirable and is received for carrying out a certain action; it is an element in the reciprocal arrangement in which parties exchange material and non-material resources. This notion of exchange, when taken in conjunction with the distinction between calculative and normative cooperation, can provide a useful framework for analysing what is going on in an organisation. Employees and employers can be said to negotiate an exchange in which they both give things and expect to receive things in return. The principal exchange will usually be wages in return for labour but other exchanges are possible, and it is worth exploring these as means of creating a climate of cooperation and commitment in an organisation. They will all, however, require the investment of material or non-material organisational resources.

ACTIVITY 5.3 . . . .

Suggest two or three benefits employees might expect to receive in exchange for participating in improvement programmes. Classify each one as calculative or normative.

Organisational culture is a narrower concept than climate, but one which overlaps with it. It is conceptualised and defined in a variety of ways but one definition that is widely accepted is that of Schein:

> [Organisational culture is] a pattern of basic assumptions – invented, discovered, or developed by a given group as it learns to cope with its problems of external adaptation and internal integration – that has worked well enough to be considered valid and, therefore, to be taught to new members as the correct way to perceive, think, and feel in relation to those problems.
>
> (Schein, 1985, p. 9)

Martin and Siehl (1990) focus on what cultures do:

> First, cultures offer an interpretation of an institution's history that members can use to decipher how they will be expected to behave in the future. Second, cultures can generate commitment to corporate values or management philosophy so that employees feel they are working for something they believe in. Third, cultures serve as organizational control mechanisms, informally approving or prohibiting some patterns of behavior. Finally, there is the possibility, as yet unsupported by conclusive evidence, that some types of organizational cultures are associated with greater productivity and profitability.
>
> (Martin and Siehl, 1990, p. 71)

Culture is sometimes used alongside another word to convey the notion of a collective ethos of a particular sort. Thus, for example, the words 'safety culture' have long been applied to those transport undertakings that were thought to prize passenger safety more highly than profit, and more recently the idea of organisations that exhibit a 'quality culture' has arisen. One difficulty here, however, is the implication that culture is a monolithic phenomenon, with a single culture being shared throughout an organisation. Some authors, for example Martin and Siehl (1990), argue that conflicting subcultures can co-exist alongside the dominant culture in an organisation. Martin and Siehl identify three different types of subculture, which they label enhancing, orthogonal and countercultural:

> An enhancing subculture would exist in an organizational enclave in which adherence to the core values of the dominant culture would be more fervent than in the rest of the organization. In an orthogonal subculture, the members would simultaneously accept the dominant culture of the organisation and a separate, unconflicting set of values particular to themselves. For example, an accounting division and a research and development (R&D) department may both endorse the values of their firm's dominant culture, while retaining separate sets of values related to their occupational identities, such as 'going by the numbers' for the accounting department and 'valuing innovation' in the R&D department.

> Some core values of a counterculture ... present a direct challenge to the core values of a dominant culture. A dominant culture and a counterculture ... exist in an uneasy symbiosis, taking opposite positions on value issues that are critically important to each of them.

> (Martin and Siehl, 1990, p. 73)

As a research student I had experience of a situation where the effects of an enhancing subculture were evident. I discovered that the people carrying out final inspection before goods left a bearings factory were implementing their own, more rigorous, standards instead of those laid down by the company. The dominant culture accepted that it was important to meet the quality standards agreed with customers, but the enhancing subculture was to try to achieve perfection. Another example can be found in Pfeffer and Coote (1991):

> A report on the treatment of post-operative pain, for example, found that many patients experienced an unacceptable degree of discomfort after surgery. The problem was that analgesia was inappropriately administered, sometimes because junior doctors did not know what to prescribe, but more often because a majority of nurses – at whose discretion relief was administered – disapproved of the use of opiates.

> (Pfeffer and Coote, 1991, pp. 8–9)

When selecting an approach, thinking about the organisation climate and the culture and any subcultures is important in two ways. The first concerns the fit between the climate and culture(s) and the approach. For example, soft systems methodology requires a high degree of participation from the 'actors' involved. Will the climate and the culture support this? Six Sigma is very rigorous in terms of the requirements to put particular structures and processes in place. Will these be welcomed and supported?

ACTIVITY 5.4. . . . .

Give two more examples like those in the paragraph above using the ISO 9000 series approach and the Systems Failures Method.

The second concerns any need there might be to develop a climate and a culture that support willing participation in problem-solving and improvement activities so that there is a fit between the new culture and the approach selected.

Let me say first of all that the notion that cultures can be shifted in a particular direction is not universally accepted. Hildebrandt et al. put the question well:

> To what degree is it possible through a conscious process to formulate a new, but expedient culture and next have it accepted in an organisation by means of a rational, centrally directed process of change?

(Hildebrandt et al., 1991, p. 4)

Reading the literature has led me to conclude that the answer to this question is by no means clear, but I'm setting these doubts aside here because the injunction to foster an appropriate culture is extremely prevalent in the problem-solving and improvement literature. It is therefore important to consider how this might be done, assuming you accept that it is 'do-able'.

One way is by undertaking what has been called 'the grand technocratic project' (Alvesson, 2002, p. 178). This is a top-down approach and consists of a series of stages along the following lines:

Step 1: Evaluating the situation of the organization and determining the goals and strategic direction.

Step 2: Analysing the existing culture and sketching a desired culture.

Step 3: Analysing the gap between what exists and what is desired.

Step 4: Developing a plan for developing the culture.

Step 5: Implementing the plan.

Step 6: Evaluating the changes, making new efforts to go further and/or engaging in measures to sustain the cultural change.

(Alvesson, 2002, p. 178)

Within this, the actual plan is likely to draw on some or all of the following:

- New recruitment and selection procedures so that people expected to be supportive of a desired culture will be hired. Sometimes this is combined with laying off and/or replacing people perceived as not being of 'true grit'.

- New forms of socialization and training programmes to signal the desired values and beliefs.

- Performance appraisal systems in which culturally correct ways of being and behaving are rewarded and encouraged.

- Promotion of people expressing and symbolizing the desired culture.

- Leadership which communicates cultural values in talk, actions and material arrangements, e.g. vision talk and for-public-consumption acts by the top manager.

- The use of organizational symbols – particular use of language (slogans, expressions, stories), actions (use of meetings in a ritual way, the visible use of managers' time to signal what is important) and material objects (corporate architecture, logotype, dress code).

(Alvesson, 2002, p. 178)

Some of these techniques can be seen in the following extract taken from a case looking at Richardson Sheffield, the 'manufacturer and marketer of a distinctive line of kitchen knives [that] was driven by Upton's [its Chairman's] belief in continuous upgrading in production and engineering, aggressive product development and focusing on the customer' (Mintzberg et al., 1998, p. 810):

> Many of Richardson's organizational norms and management practices had been deeply influenced by the leader's management style – his bias for action, attention to detail, commitment to customer service and dedication to the company had all been absorbed into the company culture. Said one executive:
>
>> Bryan [Bryan Upton, the Chairman of Richardson Sheffield] personalizes the culture of our company. In order to survive here, you have got to be very open and have a sunny personality, yet be able to work under pressure and give and take with Bryan and others. Those who are quiet and sensitive are so unhappy that they have to leave.
>
> [...]
>
> The work pressure was felt in all parts of the company. David Williams, the manufacturing systems manager, felt that Bryan Upton subjected him to time pressures and deadlines inappropriate for a development laboratory. 'I have learned to take the flak from these people and do what I can at my pace,' he said. 'But, it can

be hard when long-term development projects are turned on and off depending on the latest crisis.' Another term often used by managers to describe their work environment was 'unstructured'. Said one:

> We have an active, even unsociable culture, and people who revel in unstructured situations, and are self-motivated and pressure-driven, will do well here. When I joined the company, I was depressed that people didn't have time to meet with me. I later realized that they hold off to see if the new person is going to stay. However, once you're on board, this company works like a family. But you are not mollycoddled – you have to be self-directed. It's fun if you like living on your adrenaline.

(Mintzberg et al., 1998, pp. 813–14)

2.2 Skills, attitudes and preferences

I would be surprised if, as you studied Block 4, there were not some approaches that were more to your liking than others.

ACTIVITY 5.5. . . . .

Think back over Block 4. Which approach appealed to you most and which least? Why do you think that is the case?

Some people, for example, take naturally to the hard systems approach. They like the way it helps to clarify situations and the rigour it introduces into the evaluation and selection of options. Other people prefer something much less prescriptive such as the Excellence approach, or they would choose soft systems methodology because it provides a clear path to follow but does not make assumptions about the exact nature of the problem or what would constitute a solution. If you are just making a choice on your own behalf you will know your own preferences and there will be no problem in following them, provided that there is a good enough fit between the approach and the characteristics of the problem/opportunity and its context. However, when you are making a choice involving other people the choice can be more difficult.

The skills aspect of the third set of factors is very important but relatively easy to deal with because it is largely under the control of the decision makers. They should make sure they know the existing skills of those involved, perhaps by undertaking a skills audit if necessary, and provide training, mentoring and supervision as required in order to fill any skills gaps. However, when it comes to the attitudes and preferences of others, selection becomes more difficult. It is here that knowledge of the work of Kirton (1994) can be helpful.

Kirton developed the adaptation–innovation theory to account for the differences in individuals' thinking-style preferences. This theory holds that a continuum exists between those whose thinking-style preference is highly adaptive and those whose thinking-style preference is highly innovative, and that an individual naturally occupies a position along this continuum. The characteristics of the two extremes are shown in Table 5.1.

Table 5.1   Characteristics of adapters and innovators

The adapter	**The innovator**
Characterized by precision, reliability, efficiency, methodicalness, prudence, discipline, conformity	Seen as undisciplined, thinking, tangentially approaching tasks from unsuspected angles
Concerned with resolving residual problems thrown up by the current paradigm	Could be said to search for problems and alternative avenues of solution, cutting across current paradigms
Seeks solutions to problems in tried and understood ways	Queries problems' concomitant assumptions: manipulates problems
Reduces problems by improvement and greater efficiency, with maximum of continuity and stability	Is catalyst to settled groups, irreverent of their consensual views; seen as abrasive, creating dissonance
Seen as sound, conforming, safe, dependable	Seen as unsound, impractical: often shocks his [sic] opposite
Liable to make goals of means	In pursuit of goals treats accepted means with little regard
Seems impervious to boredom, seems able to maintain high accuracy in long spells of detailed work	Capable of detailed routine (system-maintenance) work for only short bursts
Is an authority within given structures	Tends to take control in unstructured situations
Challenges rules rarely, cautiously, when assured of strong support	Often challenges rules, has little respect for past custom
Tends to high self-doubt. Reacts to criticism by closer outward conformity. Vulnerable to social pressure and authority; complaint	Appears to have low self-doubt when generating ideas, not needing consensus to maintain certitude in face of opposition
Is essential to the functioning of the institution all the time, but occasionally needs to be 'dug out' of his system	In the institution is ideal in unscheduled crises, or better still in helping to avoid them, if he can be controlled

(Source: Kirton, 1994, p. 10)

It is essential to recognise that adaptation–innovation theory is essentially value-free; it is not saying that innovators are better or worse problem solvers than adapters. Again, it is the fit between the preferred style and the situation that is important. It is clear from Table 5.1 that adapters would be more suited to the formal application of many of the techniques for continuous improvement, whereas innovators are much more likely to deliver when radical improvement is required. In many situations it might seem as though the ideal method of working would be to combine high adapters with high innovators and so benefit from both sets of characteristics. However, as Kirton points out, the two do not readily combine:

> They often tend to irritate one another and hold pejorative views of one another. ...
>
> So innovators are generally seen by adaptors as being abrasive and insensitive, despite the former's denial of having these traits. This misunderstanding often occurs because the innovator attacks the adaptor's theories and assumptions: explicitly when he feels that the adaptor needs a push to get him out of his rut or to hurry him in the right direction; implicitly when he shows a disregard for the rules, conventions, standards of behaviour, etc. of his work group. What is even more upsetting for the adaptor is the fact that the innovator does not even seem to be aware of the havoc he is causing. Innovators may also appear abrasive to each other, since neither will show much respect for the other's theories, unless of course their two new points of view happen to coincide temporarily. Innovators perceive their work environment as more turbulent (Kirton, 1980) than do adaptors, who suspect innovators as disturbers of peace.
>
> Adaptors may also be viewed pejoratively by innovators, who feel that the more extreme adaptors are far more likely to reject them and their ideas than collaborate with them. Innovators tend to see adaptors as stuffy and unenterprising, wedded to systems, rules and norms which, however useful, are too restricting for their (the innovator's) liking.
>
> (Kirton, 1994, pp. 51–2)

One use to which Kirton's work has been put is to investigate the effects of culture (national and organisational) on problem solving. Crookes and Thomas explored 'some frequently voiced stereotypes about managers and problem solving' (1998, p. 583; Offprint 11) in Hong Kong and found that there was a difference in preferred problem-solving behaviour between expatriate and Chinese managers in the private sector and in the civil service. For problem-solving and improvement teams with members from different cultures, such as those that might be found in development projects or global organisations, these differences may be important when it comes to selecting an approach.

 Now read Offprint 11.

Having looked at choice of approach I shall now move on to a selection of topics that are important in managing problem-solving and improvement activities. The first is concerned with teams.

3 WORKING WITH PEOPLE

3.1 Understanding teams

ACTIVITY 5.6. . . . .

Many of the techniques you saw in Block 3 are designed to be used by teams and almost all the approaches in Blocks 1 and 4 make implicit or explicit use of teams. Suggest three advantages of group working in relation to problem solving and improvement.

As Grünig and Kühn point out:

> In business it is becoming increasingly common for problems to be solved by a group of people.
>
> [...]
>
> The increasing popularity of collective decision-making in business is frequently justified by arguing that it leads to better decisions.
>
> (Grünig and Kühn, 2005, pp. 197, 198)

However, they then go on to challenge the validity of this justification, saying 'whether this is true is debatable' (p. 198). They give five reasons (pp. 202–204) why what they call 'committee-based decision-making' may be less advantageous than individual decision-making. These can be summarised as follows:

1 Members of a group strive for conformity.

2 Each individual member's sense of responsibility is diminished.

3 Members' perceptions of the reality of the situation are restricted.

4 There is a higher readiness to accept risk.

5 High group cohesion can increase the motivation of individuals but it is more common for motivation to be reduced as a result of group membership.

A concept that is very relevant in thinking further about this list is 'groupthink'. This name was coined by Janis to refer to:

> a mode of thinking that people engage in when they are deeply involved in a cohesive in-group, when members' strivings for unanimity override their motivation to realistically appraise alternative courses of action.
>
> (Janis, 1972, p. 9)

He went on to say that groupthink:

> refers to a deterioration of mental efficiency, reality testing, and moral judgement that results from in-group pressures.
>
> (p. 9)

By looking at a series of American policy 'fiascos' and successes Janis identified eight symptoms of groupthink that can help to define the groupthink syndrome by observation. He has combined these to form three main types (Janis, 1982), which I have summarised in Table 5.2.

Table 5.2   Symptoms of groupthink

Type I. Overestimates of the group, its power and morality

1   An illusion of invulnerability, which creates excessive optimism and encourages the taking of extreme risks

2   Unquestioned belief in the group's inherent morality, inclining the members to ignore the ethical or moral consequences of their decisions

Type II. Closed-mindedness

3   Collective efforts to rationalise so that members discount warnings or any other information that might lead them to reconsider their assumptions before they recommit themselves to past policy decisions

4   Stereotyped views of enemy leaders as too evil, weak or stupid to counter attempts to defeat their purposes, even when these attempts are risky

Type III. Pressures toward uniformity

5   Self-censorship of deviation from the apparent group consensus, thereby minimising the importance of any self-doubt

6   Shared illusion of unanimity resulting partially from self-censorship and partially from the false assumption that silence means consent

7   Direct pressure for loyalty on members who express arguments contrary to the prevailing view

8   The emergence of 'mindguards' who protect the group from adverse information

Janis offers recommendations to avoid bad group decisions.

- Appoint group members to roles that evaluate group processes and the contributions of other members.

- In discussion, focus fully on areas of doubt and uncertainty. Be tenacious in challenging tethered assumptions and presenting the fullest possible information. Challenge the data, and assess their significance and reliability.

- Ensure that group leaders solicit and receive feedback or criticism about their judgements from other members of the group. This feedback and examination process needs to seen as a contributor to quality and not as a gripe or complaint mechanism. The potential for holding grudges and punishment of 'critics' must be avoided.

- Help the group to take regular time-out breaks to give individuals room to rethink, reformulate, gather further data and re-present. Sub-groups can do more detailed work for re-presentation. This encourages the group to cohere as a unit for analysing problems and developing solutions, as difficult tasks are decomposed and various options are properly synthesised.

Another phenomenon associated with team working that is sometimes observed is 'choice shift' (also known as group polarisation). This usually takes the form of a 'risky shift' where a team becomes more radical and more willing to take risks than individual members, but it can also appear as a 'cautious shift' where the team becomes more conservative than individuals would be. One way of guarding against these effects is to make the identification and assessment of the risks associated with their recommendations part of the group's remit; this will focus members' attention on risk and their attitudes to it.

When setting up a team you might find Tuckman's simple model of team development (Tuckman, 1965) helpful in understanding the team's dynamics. According to the model there are four basic stages in group formation: forming, storming; norming, and performing.

The forming stage is the introductory stage, during which issues about what the group wants to do and how it can operate will be raised. There will usually be a fairly rapid consensus on most issues, but it will be a false consensus, since the people in the group do not yet know each other well enough to risk saying what they really think.

The storming stage is the most difficult, and the most critical, stage of group development. It occurs when the false consensus of the forming stage starts to crack, when people start to argue, have fundamentally different ideas about how to proceed or who is the leader. The greatest error in this stage is to try to 'patch it up' or 'pretend it is not there'. These strategies leave the group stranded in the false consensus stage and they never develop real working relations. The best strategy is to find a way in which the disagreements can be safely voiced. (In some groups this is handled by humour and banter, which can serve the purpose.) It is crucial that everyone feels that their views have been heard in this stage. If they end up feeling unheard, then they will usually mentally retreat from the group and cease to make a useful contribution.

Once out of the storming stage, the group can then enter the norming stage, which creates genuine patterns of responsibility and working in the group. Tasks are effectively shared or farmed out. There is no further dispute about how to do things: it is just a question of who does what.

This moves smoothly into the fourth stage, that of performing. This last stage is when the group is functioning really effectively.

It seems to be a general rule in this whole cycle that the groups who dare to face the storming stage most honestly are the ones that end up performing the best.

3.2 Leadership

I am sure it will be no surprise to you that there is a section on leadership in this block. However, it is a huge topic and I can only scratch the surface here.

 Start by reading Offprint 12.

In the problem-solving and improvement context, leadership is particularly important at two levels: at senior level where strategic management is needed, and at team level. At the senior level, Kanji and Moura E Sá (2001, p. 709) identify what they believe are 'the critical success factors for leadership excellence':

- the existence of strong and shared organizational values (which provide the foundation for the identity of the organization and are reflected in its mission, vision, strategy and management practices);

- the development and communication of an inspiring vision;

- the definition of a mission that states what the organization stands for;

- the development of a strategy aligned to the mission and vision and able to create a sustainable competitive advantage over the competitors;

- the establishment of an organizational structure and operational mechanisms that facilitate the implementation of the mission, vision and strategy.

(Kanji and Moura E Sá, 2001, p. 709)

3.3 Gaining commitment

In this section I shall concentrate solely on theories of motivation that I think are particularly relevant when trying to find ways of fostering people's commitment to participating in and supporting improvement programmes.

The most widely known theories of motivation are concerned with the types of incentive or goal that people seek to achieve in order to be satisfied and perform well at work. They attempt to identify specific content factors in the job environment and for this reason they are known as content theories. For example, Herzberg, one of the best-known theorists, argues that job features such as responsibility and recognition motivate people to expend effort, the implication of this being that the route to higher levels of motivation is to increase opportunities to experience recognition and responsibility.

Although content theories serve as useful explanations of job satisfaction and as points of departure for redesigning jobs to increase motivation, they have been criticised for oversimplifying such a complex concept as work motivation. A second class of theories has grown up which seeks to address these criticisms. The theories in this class are known as process theories. They attempt to explain the general processes that lead to choices between different courses of action, various degrees of effort expenditure, and the persistence of this effort over time. They base their analyses on how aspects of an environment interact to produce motivation, rather than just concentrating on the content of the environment, and are thus described as being 'content free'. Unlike content theories, which assume certain universal needs, they maintain that different people are motivated by different things, that individuals are aware of what they want and that they will move towards attaining these things on a more or less rational basis. Expectancy theory belongs to this process class of theories, and I am now going to examine a basic form of this theory.

Expectancy theory holds that a person's decision to work hard on a particular task is a function of:

1 the individual's estimate that expending effort to achieve a particular goal will be followed by certain outcomes, and

2 the desirability of those outcomes to the individual.

The first of these variables is usually expressed as a probability and is known as the expectancy, and the second is termed the valence.

In more everyday language, the theory maintains that people will expend effort if they think that something desirable will actually happen as a consequence of that effort. For example, even if people think that quality is important, there would be little point in them putting in extra effort to find ways of improving it if they believed that their extra effort would make no difference. A relationship between effort and outcome is called a performance–reward contingency. It is important to note that even when this contingency exists, it has little motivating potential in itself unless the individual concerned views the outcome as desirable enough to pursue it.

Expectancy theory is often expressed as a mathematical formula:

$$\text{strength of motivation} = \Sigma \, (\text{expectancy} \times \text{valence})$$

The sigma sign (Σ) is important. Any action is likely to have a number of salient outcomes, some with positive valences and some with negative. The expression (expectancy $\times$ valence) therefore has to be calculated for each outcome and these numbers have to be summed to produce a single value for the strength of motivation.

ACTIVITY 5.7 . . . .

Three salient outcomes of volunteering to be a member of an improvement team might be: higher self-esteem, the need to learn new skills, and ridicule from workmates. Which are likely to have positive valences and which negative? Suggest two further outcomes.

The multiplication sign in the expression (expectancy × valence) is of particular significance too. If an individual is indifferent towards a particular outcome or feels that nothing that can be done would allow a particular goal to be achieved, then either the valence or the expectancy is zero. Since the strength of motivation is the former multiplied by the latter, it too will be zero.

What do performance–reward contingencies entail for participation in improvement activities? To begin with, they suggest that, unless the people involved actually feel that extra effort will result in an improvement that delivers benefits, they will not be motivated to put in that effort. This in turn suggests another requirement: employees need feedback about outcomes. For example, certain changes may ensure that customer needs are more fully met, but in motivational terms this is not going to affect the individuals concerned unless they are aware of this outcome. I shall look at the monitoring of improvement later in the block.

ACTIVITY 5.8. . . . .

Draw a diagram to show the positive feedback loop that can be generated by good improvement results.

In trying to make sure that the outcomes of an improvement exercise will be sufficiently desirable for people to exert effort to attain them, there are two approaches to consider. Either you can try to persuade people that making improvements is a worthwhile end in itself for them – and consequently worth the expenditure of effort for its own sake; or you can link the attainment of certain levels of performance to outcomes that the employees value anyway. An example of the latter would be a financial reward. Relating these two approaches to the concepts of cooperation based on calculative and normative means (as discussed in Section 2.1), then the latter strategy corresponds to calculative means, whereas the former relies on normative means.

Financial rewards can be problematic, especially if they are given to individuals rather than to groups. If you are going to use normative mechanisms you must establish the importance of improvement in the minds of everyone involved in the enterprise. This can be a formidable task because it can require a change of attitudes and values. People do not generally decide that a particular issue is of importance just because someone else tells them that they should. Many factors are involved, not least how the issue relates to their interests, whatever they perceive those interests to be. Trust, dependability and leadership from the top are all very important here.

Like content theories, process theories of motivation are not perfect, and Warr (1978) has drawn attention to some of the limitations of expectancy theory. A major drawback is that it assumes people to be rational, informed decision makers who can process complex information and who typically look some way into the future before acting. Although this view is applicable to some behaviour, it does not encompass it all by any means. Nevertheless, while we must acknowledge that expectancy theory has its limitations, it is worthy of study because it helps to untangle the issues surrounding the effort people are prepared to put into solving problems and achieving improvements.

A theory that goes some way towards addressing the criticisms made of expectancy theory is provided by Salancik (1977). It is concerned with the processes by which people may become committed to various policies and programmes. 'Committed', in the sense I am using it here, means a willingness to expend effort to ensure a programme's success and a readiness to 'stick with it'. While expectancy theory tries to explain behaviour by looking at motivation as the outcome of a rational effort–reward analysis of future events, Salancik's theory acknowledges that what has happened in the past must have a bearing on how people act in the future. Thus, commitment to a particular programme such as self-assessment or Six Sigma, or use of the Excellence Model, is not just governed by anticipated benefits, but is also affected by how much people feel they have psychologically invested in such schemes already. In a way this theory turns on its head the everyday assumption that attitudes govern behaviour and that in order to change people's behaviour it is first necessary to change their attitudes. Not only do people behave in accordance with their attitudes, they also develop and change their attitudes so that their present actions will be seen as consistent with their past behaviour. Thus, the relationship between attitudes to improvement and improvement-related actions is best conceived as circular and reciprocal, as shown in Figure 5.1.

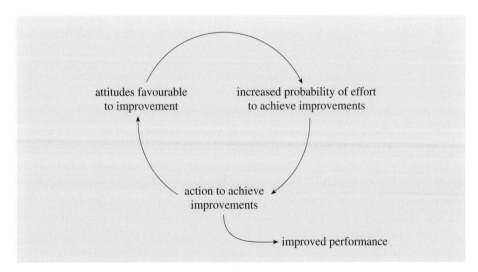

Figure 5.1   Attitudes, actions and performance

When attitudes and behaviour are conceived in this way, it becomes apparent that there are two ways of breaking into the circle. Either you can try to change attitudes by the presentation of information, advertising campaigns and so on, or you can attempt to manipulate the circumstances surrounding the introduction of a programme so that people's behaviour in the early stages encourages their commitment to the programme as a whole. The principle that underlies the second of these alternatives is that behaviour in one situation has implications for behaviour in another. If people are given the opportunity to behave in a certain way, then this initial behaviour is liable to commit them to a whole course of action as they strive to appear consistent and to live up to the expectations generated by their earlier actions. As time goes by, the more people have invested in an option the more committed to it they become as part of a self-justification process. What started as a vague interest, or a willingness to 'give it a go', may have become, in a short space of time, total commitment.

Obviously, not all actions are equally committing. The level of commitment a person feels depends on the extent to which they feel bound by their actions. According to Salancik, there are four characteristics of behaviour that cause it to be binding and therefore determine the level of the subsequent commitment to any course of action that is generated. These characteristics are: explicitness, revocability, publicity and volition.

I shall examine each one in turn and, where appropriate, see what it implies for managing the introduction of improvement programmes.

Explicitness

The explicitness or deniability of an action refers to how observable and unequivocal it is. It is obviously much easier to wriggle out of something if it is shrouded in ambiguity than if both the intention and action are clear. In practical terms this implies that all parties involved in a programme must know exactly what they are being asked to do, and the other people involved must be aware that they know.

Relating this characteristic to improvement, people's subsequent commitment to an improvement team is likely to be low unless those involved are explicit about how they view their involvement. The key point here is staking out the boundary of one's obligations. All parties should realise exactly what they have undertaken to do and their perceptions of this should be clearly understood – and shared – by the others involved. This relates to a key theme of 'felt responsibility', which I shall return to shortly. Explicitness makes the participants aware of what they and the other participants are, or are not, responsible for. The drawing up of a legal contract is a useful analogy. To minimise the risk of misunderstanding and withdrawal from the contract later on, much time is spent spelling out what is involved in a ruthlessly unambiguous way, so that no one need be in any doubt about what they have undertaken to do (unless they have not read the small print!).

Revocability

Although few actions are totally irreversible, they do differ in how easily they can be undone. Such differences will influence commitment. If a decision that is difficult to reverse is made and the person who made it has a change of mind, then they may have to choose between incurring the costs of trying to reverse the decision or sticking to a course of action that they hope may vindicate it even if they do not expect this to be the case. 'I told you so' may often be said out of relief as much as out of triumph.

This is not always as useful an idea as the others when thinking about commitment to programmes in an industrial or commercial setting because much of the decision making that is carried out in companies is undertaken by the representatives of particular interest groups – for example trade unions or staff associations – who may lose the mandate of those they represent. However, agreeing to give a particular programme a try for a specified period of time could limit its reversibility, at least for a while, and hence create commitment to it for that period.

Publicity

Depending on the context, a person's commitment to an act will be enhanced or lessened by the amount of publicity surrounding the act. Two factors combine to influence the degree to which an action will mean commitment: how much other people know about what has been done; and how important it is to the 'actor' that the 'audience' thinks well of them. One of the simplest means of committing yourself to acting in a certain way or achieving some objective, such as giving up smoking or losing weight, is to go round telling your friends – or anyone else you wish to think well of you – that you are going to do it. In order to save face you become bound to act in the way you have claimed you will. The same remarks made to the stranger standing next to you in a bus queue are unlikely to be as binding because of the different social context.

Volition

Whereas explicitness stakes out the boundary of what a person is responsible for, volition establishes how responsible that person feels. It thus refers to the extent to which someone is the 'owner' of their actions. Some of the major factors that affect how much volition is felt are:

- the perception of choice
- the presence of external demands for action, such as pressure from other people
- the extent to which the action is instrumental, that is, just a means to another end
- whether there are other people involved in the action.

The key mechanism by which volition affects subsequent commitment is that of 'felt responsibility'. The more responsible people feel for their actions, the greater will be their commitment to seeing through the consequences of those actions. If people perceive that they have a genuine choice in deciding to become members of an improvement team or a project team, then they are more likely to be committed to making it work than if the scheme is forced, or seen to be forced, on them by external pressures.

The number of other people contributing to the action may serve to increase or decrease commitment. On the one hand, diffusion of responsibility can occur as the number involved in an enterprise rises. If this is the case, a particular individual's responsibility will diminish, together with the commitment to making a scheme work, as more people become involved. On the other hand, some teams may generate considerable pressure on team members to perform in a certain way and consequently increase commitment to the achievement of the team's standard. The effects of social pressure on the one hand and diffusion of responsibility on the other make it difficult to predict the effect of other contributors on commitment. What is clear, however, is that if people feel genuinely responsible for agreeing to implement a programme, then their commitment to carry it through will be that much greater.

I have referred to the mechanism of felt responsibility twice so far. I shall now examine it a little more closely. It is a key determinant of commitment, for if people feel themselves to be 'owners' of a programme then they are more likely to be committed to working towards its success. A sense of 'ownership' and its associated felt responsibility are unlikely to be achieved unless those affected by a scheme feel that they are in that position by their own choosing. Choice can be a slippery idea to get hold of. It could be argued that someone has a choice if faced with the alternatives of participating in a programme or losing their job, for ultimately they have the freedom of choice to comply or to face the consequences of not doing so. But that person's commitment to making such a programme work successfully is hardly likely to be very high under such circumstances. In this respect consultation and participation in decision making may be seen as a mechanism for increasing felt responsibility for – and therefore commitment to – following through the consequences of decisions. Here there is a curious irony: the greater a person's participation in the decision-making process early in a programme, the more that person is trapped into ensuring the programme's success in the later stages. In other words, the greater the appearance of freedom at one stage, the greater the degree of co-option at another.

Where improvement requires the introduction of a whole new approach such as Six Sigma, commitment to its success from groups at all levels in the organisation is likely to be that much greater if the circumstances surrounding their decisions fulfil the four conditions already described: namely, that the decision should be explicit, public and difficult to reverse, and that there is real freedom of choice. However, giving people free choice and responsibility may prove to be a two-edged sword. If the choice is to be perceived as genuine, those who wish to implement such programmes must, on occasion, be seen to take no for an answer. If consultation about the proposed introduction of an approach such as Six Sigma is really a case of 'we'll listen to what you have to say as long as it doesn't interfere with what we've already decided to do', then it is unlikely to engender feelings of responsibility and commitment to the approach's success.

3.4 Decision making

It is tempting to assume that, once you have worked through a problem-solving or improvement approach to generate potential solutions, decisions about which options to implement will always be based on a rational approach to decision making. However, as Box 5.1 shows, this is unlikely to be the case, and it is important to be aware of that when managing problem-solving and improvement activities. Different people approach decision making in different ways. Even when the same data are apparently available to all, people will interpret and assimilate the data in different ways and at different speeds. Some people are very confident about weighing up a situation and making decisions; others less so. Some like to take more risks than others. Competences, such as the ability to listen to other people, also vary. Social pressures affect everyone to varying degrees and the approval or disapproval of colleagues may be more important to the decision maker than being 'right' every time. Political beliefs also vary and people will, for example, rank individual and social gains from a situation differently.

BOX 5.1   LIMITED RATIONALITY – DECISION MAKING IN THE REAL WORLD

Studies of decision making in the real world suggest that not all alternatives are known, that not all consequences are considered, and that not all preferences are evoked at the same time. Instead of considering all alternatives, decision makers typically appear to consider only a few and to look at them sequentially rather than simultaneously. Decision makers do not consider all consequences of their alternatives. They focus on some and ignore others. Relevant information about consequences is not sought, and available information is often not used. Instead of having a complete, consistent set of preferences, decision makers seem to have incomplete and inconsistent goals, not all of which are considered at the same time. The decision rules used by real decision makers seem to differ from the ones imagined by decision theory. Instead of considering 'expected values' or 'risk' as those terms are used in decision theory, they invent other criteria. Instead of calculating the 'best possible' action, they search for an action that is 'good enough'.

As a result of such observations, doubts about the empirical validity and usefulness of the pure theory of rational choice have been characteristic of students of actual decision processes for many years. Rational choice theories have adapted to such observations gradually by introducing the idea that rationality is limited. The core notion of limited rationality is that individuals are intendedly rational. Although decision makers try to be rational, they are constrained by limited cognitive capabilities and incomplete information, and thus their actions may be less completely rational in spite of their best intentions and efforts.

In recent years, ideas of limited (or bounded) rationality have become sufficiently integrated into conventional theories of rational choice to make limited rationality viewpoints generally accepted. They have come to dominate most theories of individual decision making. They have been used to develop behavioral and evolutionary theories of the firm. They have been used as part of the basis for theories of transaction cost economics and game theoretic, information, and organizational economics. They have been applied to decision making in political, educational, and military contexts.

Information constraints

Decision makers face serious limitations in attention, memory, comprehension, and communication. Most students of individual decision making seem to allude to some more or less obvious biological constraints on human information processing, although the limits are rarely argued from a strict biological basis. In a similar way, students of organizational decision making assume some more or less obvious information constraints imposed by methods of organizing diverse individuals:

1   *Problems of attention* Time and capabilities for attention are limited. Not everything can be attended to at once. Too many signals are received. Too many things are relevant to a decision. Because of those limitations, theories of decision making are often better described as theories of attention or search than as theories of choice. They are concerned with the way in which scarce attention is allocated.

2   *Problems of memory* The capabilities of individuals and organizations to store information are limited. Memories are faulty. Records are not kept. Histories are not recorded. Even more limited are individual and organizational abilities to retrieve information that has been stored. Previously learned lessons are not reliably retrieved at appropriate times. Knowledge stored in one part of an organization cannot be used easily by another part.

3   *Problems of comprehension* Decision makers have limited capacities for comprehension. They have difficulty organizing, summarizing, and using information to form inferences about the causal connections of events and about relevant features of the world. They often have relevant information but fail to see its relevance. They make unwarranted inferences from information, or fail to connect different parts of the information available to them to form a coherent interpretation.

4   *Problems of communication* There are limited capacities for communicating information, for sharing complex and specialized information. Division of labor facilitates mobilization and utilization of specialized talents, but it also encourages differentiation of knowledge, competence, and language. It is difficult to communicate across cultures, across generations, or across professional specialties. Different groups of people use different frameworks for simplifying the world.

As decision makers struggle with these limitations, they develop procedures that maintain the basic framework of rational choice but modify it to accommodate the difficulties. These procedures form the core of theories of limited rationality.

(Source: March, 1994, pp. 160–2)

Table 5.3 shows some common methods of decision making and the circumstances in which each can be helpful. Two important factors in deciding which of the methods to use are the time taken to reach a decision and the degree of ownership of those making the decision, as shown in Figure 5.2. Of course the ideal would always be low time and high ownership, that is, consensus.

Table 5.3   Common methods of decision making

Some common approaches	When to use each approach
Consensus	On strategically important issues when there is ample time to reach agreement
Autocratic (unilateral or top-down)	Important, time-sensitive matters such as a safety issue, or less important operational issues that have been delegated to a single decision-maker or a committee
Autocratic plus consultation	The same as autocratic, when more ownership is desired and there is insufficient time to produce a consensus
Majority voting	Important issues in which there is insufficient time to reach consensus, and when the ownership of most participants is desirable, as in the case of a democratic election. This is also a useful 'fallback' to consensus. [It is worth noting that it can be more effective to start by voting options out.]
Third party	When there is an important issue that is likely to get stuck at impasse, and when some commonly held standard of fairness is desirable, as in the case of a judge, jury, or arbitrator
Luck/chance	On less important issues when fairness is more important than ownership of the decision, as in drawing straws or tossing a coin

(Source: Glaser, 2005, p. 153)

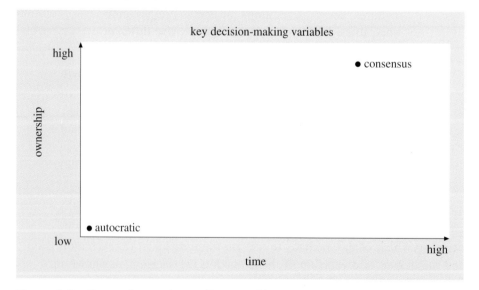

Figure 5.2   Ownership and time (Source: Glaser, 2005, p. 150)

ACTIVITY 5.9 . . . .

Plot the other methods shown in Table 5.3 on Figure 5.2.

4 MONITORING PROBLEM SOLVING AND IMPROVEMENT

In order to manage any process it is necessary to set objectives, measure the extent to which they are being met, and then plan and execute actions that will make good any shortfall and improve future performance. This applies just as much to problem solving and improvement as to any other process.

ACTIVITY 5.10 . . . .

Suggest two further reasons why it is important to monitor problem solving and improvement.

In Programme 3 you heard that SottishPower's Six Sigma programme costs £4.5 million per year and delivers an annual saving of £15–20 million. As Box 5.2 shows, other organisations quantify their interventions in financial terms, either as sums of money or as financial ratios, and they do so with varying degrees of precision.

BOX 5.2 REPORTED RESULTS

Example 1

A leading asset management house recently asked Horizon to review their purchasing. Our brief was to improve quality, maximise environmental efficiency and optimise costs where possible. £4 million was delivered without any interruption to the core business.

(Source: Horizon Associates, 2007)

Example 2

Savings were delivered through a series of solutions from the Business Transformation (BT) programme. It should be noted that information provided relates to actual date for 2005/06 up to 30.09.05, with projections up to 31.03.06. Now nearing completion of the final stage of the programme, most of the decisions and actions have been taken and savings based on those earlier actions are now being realised in the following area:

Cost Savings = £2.328M

These are the net savings in costs arising from the implementation of BT solutions.

(Source: Scottish Executive, 2006)

Example 3

The company completed 28 Six Sigma projects during the second quarter of 2006 for a total annualised benefit of USD8.2m at an average savings value of USD293,000 per project.

(Source: M2 Communications, 2006)

Example 4

We have established a strong track record of driving efficiency improvements and I am pleased that in 2006 we improved our cost:income ratio to 50.8 per cent, from 52.8 per cent in 2005. This was achieved by our continued commitment to a range of quality improvement programmes such as lean manufacturing, which enable us to enhance the service we deliver to our customers at a lower cost. We have extended our Groupwide efficiency programme that is also allowing us to structurally reduce our cost base. As we continue to improve our efficiency and effectiveness, we are creating additional capacity for further investment to support our future growth plans.

(Source: Lloyds TSB Group, 2007)

Example 5

The mean cost per patient for acetaminophen analgesia decreased from £14 before the intervention to £6 after the intervention. We projected the annual cost reduction to be £15 100. We estimated the cost of the intervention to be £970. Thus, the cost of the intervention was recuperated within three weeks.

(Source: Ripouteau et al., 2000)

Oakland and Marosszeky (2006) have drawn up a list of the essential requirements for measuring performance. Measures must be:

1   Transparent – understood by all the people being measured
2   Non-controversial – accepted by the individuals concerned
3   Internally consistent – compatible with the rewards and recognition systems
4   Objective – designed to offer minimal opportunity for manipulation
5   Motivational – trigger a response to improve outcomes.

(Oakland and Marosszeky, 2006, p. 114)

ACTIVITY 5.11 ....

Suggest another essential requirement for measures.

Oakland and Marosszeky also give a list of problems they have seen in systems that have frustrated improvement efforts. They observe that such systems:

1   Produce irrelevant or misleading information.

2   Track performance in single, isolated dimensions.

3   Generate financial measures too late, e.g. quarterly, for mid-course corrections or remedial action.

4   Do not take account of the customer perspective, both internal and external.

5   Distort management's understanding of how effective the organization has been in implementing its strategy.

6   Promote behaviour and undermine the achievement of the strategic objectives [sic].

(Oakland and Marosszeky, 2006, p. 111)

It is clear that expressing the costs of problems and the consequences of problem solving and improvement in monetary terms can meet most (and possibly, sometimes, all) of the requirements and avoid most (and possibly, sometimes, all) of the problems, but expressing everything in monetary terms brings its own problems. For a start, some variables such as reputation in the marketplace are impossible to value and there is plenty of evidence that accurate information about things that can be valued is not always available.

Other measures that can be used to carry out 'before' and 'after' comparisons include effectiveness, efficiency and productivity. Effectiveness is concerned with the output side of a process. It is expressed as a percentage and is calculated thus:

$$\text{effectiveness} = \frac{\text{actual output}}{\text{expected output}} \times 100 \text{ per cent}$$

Efficiency is concerned with the resources actually used compared with the resources that were planned to be used, so is looking at inputs. It too is a percentage:

$$\text{efficiency} = \frac{\text{resources actually used}}{\text{resources planned to be used}} \times 100 \text{ per cent}$$

Efficiency is often calculated for a number of separate variables to give measures such as labour efficiency, materials efficiency, equipment efficiency, and so on.

Productivity measures relate outputs to inputs:

$$\text{productivity} = \frac{\text{outputs}}{\text{inputs}}$$

If you are undertaking a quality improvement project focusing on a particular process, one approach that can be used to monitor the success (or otherwise) of the project is that set out in BS 6143-1:1992. This is based on a process cost model. Construction of the process's cost model begins by modelling it in terms of:

- its name
- its outputs and customers
- its inputs and suppliers
- the controls that define, regulate and/or influence it
- its resources.

This information is presented in a simple block diagram format as shown in Figure 5.3. Figure 5.4 shows a process model of a typical manufacturing department process.

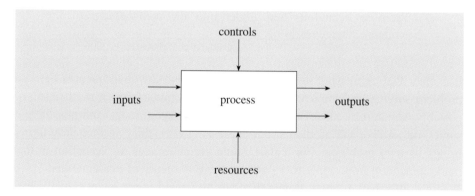

Figure 5.3   The basic process model (Source: BS 6143-1:1992, p. 4)

ACTIVITY 5.12 ....

Draw a process model for a house construction process.

The next stage in building the process cost model is to identify the key activities that are carried out in the process and determine their cost elements. Each cost element can be classified as either a cost of conformance (COC) or a cost of nonconformance (CONC). COC is defined as 'the intrinsic cost of providing products or services to declared standards by a given, specified process in a fully effective manner'. CONC is 'the cost of wasted time, materials and capacity (resources) associated with a process in the receipt, production, dispatch and correction of unsatisfactory goods and services' (BS 6143-1:1992, p. 3).

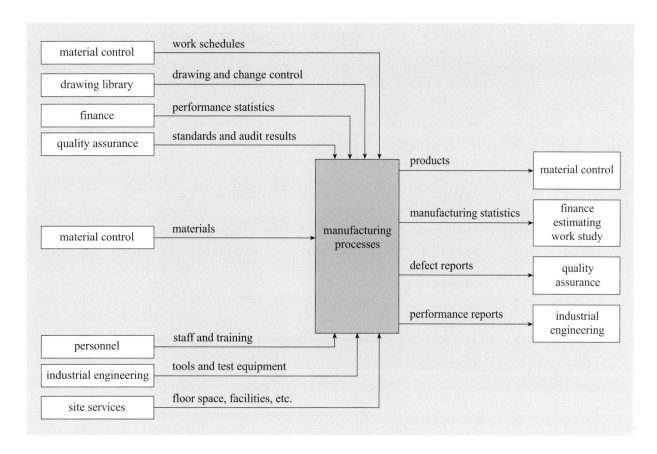

Figure 5.4   A process model of a typical manufacturing department process (Source: BS 6143-1:1992, p. 6)

ACTIVITY 5.13 ·

Quality improvements are very likely to reduce the CONC. Will the COC represent the minimum quality cost associated with a process or might it, too, offer scope for reduction?

As shown in Table 5.4, a process cost model is merely the tabulation of the information that has just been described.

Table 5.4   Part of a process cost model for a manufacturing department

Key activity	Cost of conformance	Cost of nonconformance
Planning, production engineering work study, cost control, materials and process laboratory	Part cost	Part cost (effect of engineering change, planning errors, etc.)
Production inspection and test costs	'Good' hours booked	Reinspection/retest/fault finding
Test gear depreciation, calibration and preventative maintenance	Total cost	
Breakdowns		Total cost
Production costs	'Good' hours booked	Rework
Material costs	Estimated cost	Scrap cost, overspend
Waiting time		Total cost
Cost of work held due to shortages		Total cost

(Source: BS 6143-1:1992, p. 9)

At the macro level you might expect that successful problem solving and the existence of a successful improvement programme will be reflected in a company's 'bottom line'. However, that is not always the case because the reasons why a certain level of profitability is achieved are always multi-factorial. As you may recall from Block 4, the organisation that developed Six Sigma was Motorola. It is reported that 'Motorola saved $1 billion in three years' (*Strategic Direction*, 2002, p. 8), so you might be surprised to read the following extract:

> Let's consider how Motorola is performing today after using the Six Sigma approach for more than 10 years. Motorola's market share in its main business, wireless phones, has steadily decayed in the past decade. The company's president, Bob Growney, said in February 2001 that Motorola plans to cut costs by closing, or selling, at least four of its 55 manufacturing plants – maybe as many as seven.
>
> Motorola completely misjudged the transition from analog to digital phones. Six Sigma did not help marketing do its job better. Motorola once owned the wireless phone market. Today, Nokia has 31 percent of the market, and Motorola has 15 percent, according to research done by Dataquest. Nokia's tiny high-end phones serve as today's benchmark. Additionally, Motorola's phones don't make use of many interchangeable parts, which makes them costly to manufacture. 'Clearly they need to take quite a lot of cost out of their system,' says Vivian Mamelak, an analyst with Arnhold and S. Bleichroeder. It looks like Six Sigma

didn't work in product engineering, manufacturing engineering or quality.

As a former quality professional, I realize we can't completely blame the Six Sigma program for Motorola's performance. But, likewise, we can't give Six Sigma total credit for the company's success. Motorola's problem may be that it has dropped the Six Sigma approach or could only apply it to manufacturing and, to a very limited extent, in the support areas. But management doesn't drop programs that are having a significant positive impact on the bottom line, particularly after the company has invested as much money as Motorola invested in Six Sigma.

Now, as a COO [Chief Operating Officer] of an organization that's in the process of going public, I have to look at the long-term results. The company that invented and used Six Sigma for the longest period of time has continuously decreased market share and lost technological leadership.

(Harrington, 2001)

5   PROJECT AND PORTFOLIO MANAGEMENT

Where improvement projects are concerned, good management is important both at the level of each individual project and also at the level of the overall programme. Project and portfolio management is a course-length topic in itself, but I should like to remind you here of the important elements. Figure 5.5 identifies the key aspects of the management of a single project. Fortunately, many of the tools and techniques that play a part in solving problems and delivering improvements are just as useful in the management of the project itself. For example, decision trees and risk assessment, which appeared in Section 12 of Block 3, are both recommended for use by project managers generally.

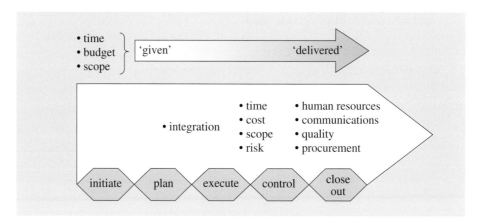

Figure 5.5   Project management (Source: Morris and Pinto, 2004, p. xvi)

The key aspects of portfolio management are shown in Figure 5.6. The biggest differences between the management of a single project and the concurrent management of a collection of projects are the need to select the projects that will become part of the portfolio, to coordinate and prioritise between projects, and to manage shared resources.

Project selection is important and can be problematic. Archer and Ghasemzadeh (1999) classify the techniques available for portfolio selection thus:

- *Ad hoc approaches* such as (a) Profiles, a crude form of scoring model, where limits are set for the various attribute levels of a project, and any projects which fail to meet these limits are eliminated (study of the human–computer interface aspects of such approaches have shown that users prefer these minimum effort approaches, whether or not they give an optimal solution), and (b) Interactive selection, involving an interactive and iterative process between project champions

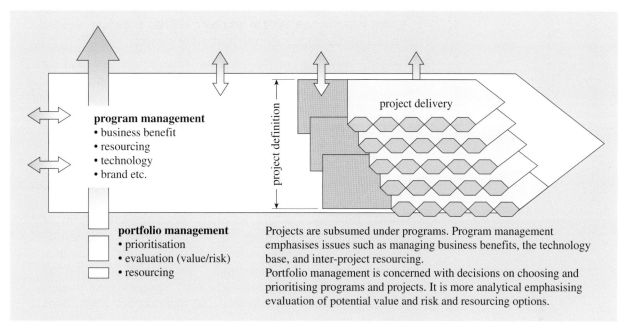

Figure 5.6   Portfolio management (Source: Morris and Pinto, 2004, p. xx)

and responsible decision maker(s) until a choice of the best projects is made.

- *Comparative approaches* include Q-Sort, pairwise comparison, the Analytic Hierarchy Procedure (AHP), dollar metric, standard gamble, and successive comparison. Q-Sort is the most adaptable of these in achieving group consensus. In these methods, first the weights of different objectives are determined, then alternatives are compared on the basis of their contributions to these objectives, and finally a set of project benefit measures is computed. Once the projects have been arranged on a comparative scale, the decision maker(s) can proceed from the top of the list, selecting projects until available resources are exhausted. With these techniques, both quantitative and qualitative and/or judgment criteria can be considered. A major disadvantage of Q-sort, pairwise comparison and AHP is the large number of comparisons involved, making them difficult to use for comparing large numbers of projects. Also, any time a project is added or deleted from the list, the process must be repeated.

- *Scoring models* use a relatively small number of decision criteria, such as cost, work force availability, probability of technical success, etc., to specify project desirability. The merit of each project is determined with respect to each criterion. Scores are then combined (when different weights are used for each criterion, the technique is called 'Weighted Factor Scoring') to yield an overall benefit measure for each project.

A major advantage is that projects can be added or deleted without re-calculating the merit of other projects.

- *Portfolio matrices* can be used as strategic decision making tools. They can also be used to prioritize and allocate resources among competing projects. This technique relies on graphical representations of the projects under consideration, on two dimensions such as the likelihood of success and expected economic value. This allows a representative mix of projects on the dimensions represented to be selected.

- *Optimization models* select from the list of candidate projects a set that provides maximum benefit (e.g. maximum net present value). These models are generally based on some form of mathematical programming, to support the optimization process and to include project interactions such as resource dependencies and constraints, technical and market interactions, or program considerations. Some of these models also support sensitivity analysis, but most do not seem to be used extensively in practice. Probable reasons for disuse include the need to collect large amounts of input data, the inability of most such models to include risk considerations, and model complexity. Optimization models may also be used with other approaches which calculate project benefit values. For example, 0-1 integer linear programming can be used in conjunction with AHP to handle qualitative measures and multiple objectives, while applying resource utilization, project interaction, and other constraints.

(Archer and Ghasemzadeh, 1999, p. 210)

Archer and Ghasemzadeh have devised a project portfolio selection framework of their own. This is shown in Figure 5.7. Although it was not developed with any particular types of project in mind – it could, for example, be used to develop a portfolio of property development projects or product design projects – its inclusion of methodology selection makes it highly relevant to improvement projects.

ACTIVITY 5.14 ....

Are there any aspects of Archer and Ghasemzadeh's framework for project portfolio selection that might be considered irrelevant where improvement projects are concerned?

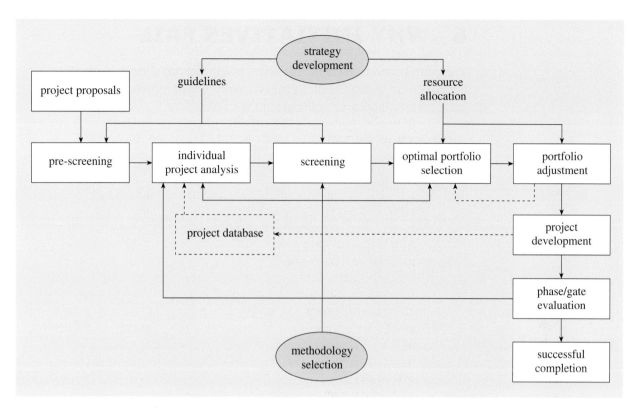

Figure 5.7   Framework for project portfolio selection (Source: Archer and Ghasemzadeh, 1999, p. 211)

6   WHY INITIATIVES FAIL

There are bound to be some instances when individual improvement initiatives are less successful than originally hoped, or when effective and practicable solutions to problems are not found.

ACTIVITY 5.15 . . . .

Suggest three reasons why effective and practicable solutions to problems might not be found.

Much more serious is the threat that a whole improvement programme might fail.

 Now read Offprint 13.

7 CONCLUSION

Before you embark on your project I would like you to finish this part of the course by watching the last two programmes on your course DVD. Each of the programmes has something to teach you in its own right about the management of problem solving and improvement and, in particular, the involvement of people, but in combination they provide an extremely good illustration of '*plus ça change, plus c'est la même chose*' (the more things change, the more they stay the same). The programmes were made twenty years apart but it is only really the internal shots of the pre-electronic trading floor of the Stock Exchange that reveal the age of Programme 5. The message I would like you to take away from this is that, over time, small parts of this course will be dismissed as fads that have 'passed' and the vocabulary in other parts will change, but the vast majority of the course has a permanence because it is based on method, rigour, analysis and sound theory. When the next 'new' thing comes along, be sceptical and look for the weaknesses but do remember that a gimmick is a problem only if there is little or no substance behind it.

You should now watch Programme 4 Problem solving in action: transformation at COSi and Programme 5 Customers, quality, competition.

Box 5.3 explains the 5S approach that is mentioned in Programme 4.

BOX 5.3   THE 5S APPROACH IN PROGRAMME 4

The 5S approach is not included elsewhere in the course because it is not really on a par with the other approaches that are covered. However, it is described briefly here in case you are not familiar with it. It is sometimes known as the 5S housekeeping approach and recommended for use in lean manufacturing. Each S originates from a Japanese word. An English set of Ss has been devised which, even though all but one are not straight translations, try to reflect the spirit of the approach (some more successfully than others). The headings below show the English translation of the Japanese as well as both sets of Ss.

Seiri, tidiness, sort. Identify the best physical organisation of the workplace bearing in mind factors such as the frequency with which items are used, items that are shared between users, heath and safety hazards, ease of movement, and the like. Discard all unnecessary items.

Seiton, orderliness, set. Implement the changes needed to achieve the best physical organisation.

Seiso, cleanliness, shine. The word *seiso* is best translated as cleanliness. Establish a programme to clean up the work environment and keep it that way by identifying who is responsible for each aspect and devising a cleaning and maintenance rota.

Seiketsu, standardisation, standardisation. Iterate the first three Ss to ensure that the optimum level of organisation and cleanliness is maintained. External (i.e. outside the area) assessment is also part of this stage of the approach.

Shitsuke, discipline, sustain. Train and motivate to build wider commitment to the approach.

You have now finished the taught component of the course and will shortly be moving on to the project. I hope you have enjoyed studying the materials and I wish you luck with your project.

REFERENCES

Alvesson, M. (2002) *Understanding Organizational Culture*, London, Sage.

Archer, N. P. and Ghasemzadeh, F. (1999) 'Project portfolio selection', *International Journal of Project Management*, Vol. 17, no. 4, pp. 207–16.

Bowey, A. M. and Carlisle, B. (1979) *Group Working*, Bradford, MCB Publications.

BS 6143-1:1992 *Guide to the Economics of Quality. Process Cost Model*, London, British Standards Institution.

Crookes, D. and Thomas, I. (1998) 'Problem solving and culture – exploring some stereotypes', *Journal of Management Development*, Vol. 17, no. 8, pp. 583–91.

Dale, B. G. and Hayward, S. (1984) *A Study of Quality Circle Failures*, Manchester, UMIST.

Davenport, T. H. and Prusak, L. with Wilson, H. J. (2003a) 'Reengineering revisited – what went wrong with the business-process reengineering fad. And will it come back?', *Computerworld*, 23 June, http://www.computerworld.com/managementtopics/management/story/0,10801,82290,00.html (accessed 4 July 2007).

Davenport, T. H. and Prusak, L. with Wilson, H. J. (2003b) *What's the Big Idea: Creating and Capitalizing on the Best Management Thinking*, Harvard, Harvard Business School Press.

Glaser, J. (2005) *Leading through Collaboration*, Thousand Oaks, Calif., Corwin Press.

Grünig, R. and Kühn, R. (2005) *Successful Decision-Making: a Systematic Approach to Complex Problems*, Berlin, Springer.

Harrington, H. J. (2001) 'Six Sigma's long-term impact', *Quality Digest*, June, http://www.qualitydigest.com/june01/html/harrington.html (accessed 31 August 2007).

Hildebrandt, S., Kristensen, K., Kanji, G. and Dahlgaard, J. J. (1991) 'Quality culture and TQM', *Total Quality Management,* Vol. 2, no. 1, pp. 1–15.

Horizon Associates (2007) *Latest News: Cost Savings Coupled with Quality Improvement*, http://www.purchasing-consultants.co.uk/news/detail.php?id = 2 (accessed 30 August 2007).

Janis, I. (1972) *Victims of Groupthink*, Boston, Houghton Mifflin.

Janis, I. (1982) *Groupthink: Psychological Studies of Policy Decisions and Fiascos*, Boston, Houghton Mifflin.

Kanji, G. K. and Moura E Sá, P. (2001) 'Measuring leadership excellence', *Total Quality Management*, Vol. 12, no. 6, pp. 701–18.

Katz, D. and Kahn, R. L. (1978) *The Social Psychology of Organizations*, Chichester, Wiley.

Kirton, M. (1994) (ed.) *Adaptors and Innovators*, revised edn, London, Routledge.

Kirton, M. J. (1980) 'Adaptors and innovators in organizations', *Human Relations*, Vol. 3, pp. 213–24.

Lloyds TSB Group (2007) *Lloyds TSB Group PLC – 2006 Final Results*, Lloyds TSB Group PLC.

March, J. (1994) 'Limited rationality' in Billsberry, J. (ed.) (1996) *The Effective Manager*, London, Sage/Open University.

M2 Communications (2006) 'Falconbridge Limited announces results for second quarter 2006', *M2 Equitybites*, 25 July, http://lxisnexis.com/uk/business/delivery (accessed 23 July 2007).

Martin, J. and Siehl, C. (1990) 'Organizational culture and counterculture: an uneasy symbiosis', in Sypher, B. D. (ed.) *Case Studies in Organizational Communication*, New York, Guildford Press.

Mintzberg, H., Quinn, J. B. and Ghoshal, S. (1998) *The Strategy Process*, Harlow, Pearson.

Morris, P. and Pinto, J. (2004) 'Introduction' in Morris, P. W. G. and Pinto, J. (eds) *The Wiley Guide to Managing Projects*, Chichester, Wiley.

Oakland, J. and Marosszeky, M. (2006) *Total Quality in the Construction Supply Chain*, Oxford, Butterworth-Heinemann.

Pfeffer, N. and Coote, A. (1991) *Is Quality Good for You?*, Social Policy Paper No. 5, London, Institute of Public Policy Research.

Ripouteau, C., Conort, O., Lamas, J. P., Auleley, G.-R., Hazebroucq, G. and Durieux, P. (2000) 'Effect of multifaceted intervention promoting early switch from intravenous to oral acetaminophen for postoperative pain: controlled, prospective, before and after study', *British Medical Journal*, no. 321, 9 December, pp. 1460–3, http://www.bmj.com/cgi/content/full/321/7274/1460 (accessed 30 August 2007).

Salancik, G. R. (1977) 'Commitment and the control of organisational behaviour and belief', in Staw, B. M. and Salancik, G. R. (eds) *New Directions in Organisational Behaviour*, Chicago, St Clair Press.

Schein, E. (1985) *Organizational Culture and Leadership*, San Francisco, Jossey-Bass.

Scottish Executive (2006) *Efficiency Technical Notes – March 2006: Part 6 Enterprise and Lifelong Learning*, http://www.scotland.gov.uk/Publications/2006/03/31095821/6/Q/Zoom/80 (accessed 30 August 2007).

Strategic Direction (2002) 'Black belts save Motorola a billion', *Strategic Direction*, Vol. 18, no. 1, pp. 8–9.

Tang, S. L., Aoieong, R. T. and Ahmed, S. M. (2004) 'The use of Process Cost Model (PCM) for measuring quality costs of construction projects: model testing', *Construction Management and Economics*, Vol. 22, no. 3, pp. 263–75.

Tuckman, B. W. (1965) 'Developmental sequence in small groups', *Psychological Bulletin,* no. 63, pp. 384–99.

Warr, P. B. (1978) 'Attitudes, actions and motives', in Warr, P. B. (ed.) *Psychology at Work*, Harmondsworth, Penguin.

ANSWERS TO ACTIVITIES

Activity 5.1

Possible reasons include:

- recommendation from another organisation
- proposal by external consultant
- attendance at an event
- information from professional or industrial association.

Activity 5.2

It immediately made me think of Box 4.10 (in Block 4) on re-engineering, especially the following:

> But in late 1996, a front-page *Wall Street Journal* article featured a confession by Hammer: 'Dr. Hammer points out a flaw: He and the other leaders of the $4.7 billion re-engineering industry forgot about people. "I wasn't smart enough about that," he says. "I was reflecting my engineering background and was insufficiently appreciative of the human dimension. I've learned that's critical."'
>
> Hammer's earlier rhetoric certainly neglected the human element, with phrases such as, 'In reengineering, we carry the wounded and shoot the stragglers', and, 'It's basically taking an ax and a machine gun to your existing organization.' This rhetoric not only made employees fear for their livelihoods; it also raised expectations of managers for revolutionary changes that couldn't be delivered.
>
> (Davenport et al., 2003a, excerpted from Davenport et al., 2003b)

Activity 5.3

My suggestions are:

- More variety in work activities. Normative.
- Resulting improvements to their own jobs that will make their work easier, more interesting, more fulfilling, or whatever. Mainly normative but could be calculative if they are paid by results.
- Recognition that their contributions are valued. Normative.
- Improved chances of advancing in the organisation. Calculative.
- Opportunity to develop portable problem-solving and improvement skills. Calculative.

Activity 5.4

The ISO 9000 series approach is somewhat bureaucratic and requires a certain degree of formality to exist. Will the climate and the culture allow and support this?

The Systems Failures Method requires openness about failures that may have occurred and avoidance of 'the blame game'. Will the climate and the culture allow this? Does a culture of blame exist?

Activity 5.6

Advantages of teams include the following:

- Teams may be representative of the stakeholders as a whole.
- Greater diversity can enrich creativity and lead to more ideas being generated.
- There is a greater pool of skills on which to draw.
- Other group members can be used as a sounding board for ideas.

Activity 5.7

Positive valence: higher self-esteem.

Negative valence: ridicule from workmates.

Either positive or negative: need to learn new skills.

Other outcomes might be:

- praise from superior
- financial reward
- improvements that make own job easier.

Activity 5.8

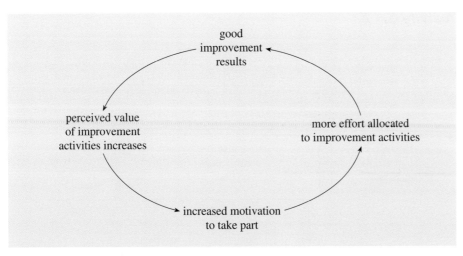

Figure 5.8   Positive feedback loop generated by 'good improvement results'

Activity 5.9

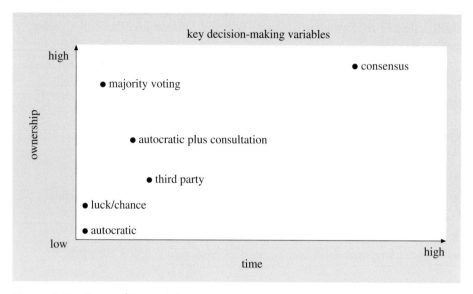

Figure 5.9   Ownership and time

Activity 5.10

One reason is the consequences monitoring may have for the motivation of those involved. A second reason is the ability to monitor the returns on investment (ROI) in problem solving and improvement.

Activity 5.11

I suggest that measures must also be:

- Relevant – do not give rise to or encourage sub-optimal behaviours.

Activity 5.12

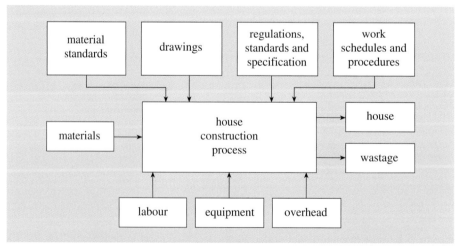

Figure 5.10   A process model for a house construction process (Source: adapted from Tang et al., 2004, p. 264)

Activity 5.13

The COC is likely to offer scope for improvement because the process is taken as a given in determining it. If the design of the process can be improved, it should be possible to reduce the COC.

Activity 5.14

I suggest that many people would consider the notion of 'portfolio completion' irrelevant where improvement projects are concerned.

Activity 5.15

Reasons include the following:

- Historical data are not available to carry out the analysis required.
- There is a failure of creativity on the part of those working on the problem.
- The solution is not practicable because of the high level of investment required.
- The solution is not practicable because implementation would generate problems elsewhere.
- Conflict in the problem situation cannot be resolved.

ACKNOWLEDGEMENTS

Grateful acknowledgement is made to the following sources:

Text

Oakland J., Marousszeky M., (2006), 'Performance Measurement and the Improvement Cycle', *Total Quality in the Construction Supply Class*, Elsevier; March J., & Billsberrry J., (1995), '21 Limited Rationality', *The Effective Manager - Perspectives And Illustrations*, Sage Publications Limited.

Tables

Table 5.1: Kirton M., (1994), 'J.A. Theory of Cognitive Style & 3 Adaptors and Innovators at Work', *Adaptors and Innovators*, Taylor Francis, Routledge, Garland.; Table 5.4: Permission to reproduce extracts from British Standards is granted by BSI. British Standards can be obtained in PDF format from the BSI online shop: http://www.bsi-global.com/en/Shop/ or by contacting BSI Customer Services for hardcopies: Tel: +44 (0)20 8996 9001, email: mailto:cservices@bsi-global.com.

Figures

Figure 5.2: Glazer J., (2005), 'Reaching Effective Agreements', *Leading Through Collaboration*, Corwin Press Inc. and Sage Publications Limited.; Figures 5.3 and 5.4: Permission to reproduce extracts from British Standards is granted by BSI. British Standards can be obtained in PDF format from the BSI online shop: http://www.bsi-global.com/en/Shop/ or by contacting BSI Customer Services for hardcopies: Tel: +44 (0)20 8996 9001, email: mailto:cservices@bsi-global.com; Figures 5.5 and 5.6: Morris P.W.G., and Pinto J., (2004), *The Wiley Guide to Managing Projects*, John Wiley & Sons Limited; Figure 5.7: Archer N.P. and Ghasemzadeh F., (1999), *International Journal of Project Management,* Elsevier Science; Figure 5.10: Tang S.L., Aoelong R. T., Ahmed S.M., (2004), 'Introduction', *Construction Management And Economics,* Taylor & Francis Limited.

Index of techniques

Techniques

In the list of techniques below the bold numbers refer to blocks and the other numbers to sections within blocks.

A

Activity sequence diagrams **3**:6.1

Affinity diagrams **3**:3.1

Arrow diagrams **3**:3.6

Attributes charts **3**:11.3

B

Balanced scorecard **3**:13.3

Bar charts **2**:2.2

Benchmarking **3**:9.4

Boxplots **2**:3.5

Brainstorming **3**:8.1

Brainwriting **3**:8.2

C

Cause-and-effect diagrams **3**:2.2

Cognitive mapping **3**:6.8

Conceptual models **4**:5.1

Control charts for attributes **3**:11.3

Control charts for variables **3**:11.3

Creative problem solving (CPS) **3**:8.5

Creativity techniques **3**:8; **3**:Appendix

Cumulative frequency diagrams **2**:2.3

Cumulative sum charts (cusums) **3**:11.3

Customer/business impact analysis **4**:7.1

D

Decision trees **3**:12.2

E

Environmental scanning **3**:9.3

Exponentially weighted moving average (EWMA) **3**:11.3

EWMA charts **3**:11.3